Second-Order Nonlinear Optical Characterization Techniques

An Introduction

Second-Order Nonlinear Optical Characterization Techniques

An Introduction

Thierry Verbiest
Koen Clays
Vincent Rodriguez

CRC Press
Taylor & Francis Group
Boca Raton   London   New York

CRC Press is an imprint of the
Taylor & Francis Group, an **informa** business

CRC Press
Taylor & Francis Group
6000 Broken Sound Parkway NW, Suite 300
Boca Raton, FL 33487-2742

First issued in paperback 2017

ISBN 13: 978-1-138-11826-3 (pbk)
ISBN 13: 978-1-4200-7071-2 (hbk)

Library of Congress Cataloging-in-Publication Data

Verbiest, Thierry.
  Second-order nonlinear optical characterization techniques : an introduction / Thierry Verbiest, Koen Clays, Vincent Rodriguez.
      p. cm.
  Includes bibliographical references and index.
  ISBN 978-1-4200-7071-2 (alk. paper)
    1. Materials--Optical properties. 2. Materials--Analysis--Technique. 3. Surfaces (Technology)--Analysis--Technique. 4. Molecular structure--Optical aspects. 5. Optical measurements--Technique. 6. Nonlinear optics. 7. Second harmonic generation. I. Clays, Koen. II. Rodriguez, Vincent. III. Title.

TA417.43.V47 2009
620.1'1295--dc22   2008050889

Visit the Taylor & Francis Web site at
http://www.taylorandfrancis.com

and the CRC Press Web site at
http://www.crcpress.com

Contents

Preface

This book focuses on characterization techniques that are based on second-order nonlinear optical processes. Examples of second-order nonlinear optical processes include the linear electro-optic (or Pockels) effect and harmonic generation processes such as second-harmonic generation (SHG) and sum- and difference-frequency generation (SFG and DFG). The first effect has been used mainly in applications such as ultra-fast modulators and switches, whereas harmonic generation processes have been applied in frequency conversion devices. Furthermore, second-harmonic and sum-frequency generation received attention because of their potential as powerful characterization techniques in physical and physico-chemical labs.

Second-harmonic generation or frequency doubling was first observed in 1961 by Franken and coworkers at the University of Michigan. They focused a ruby laser (wavelength 694 nm) onto a quartz sample and sent the output through a spectrometer. The spectrum was recorded on photographic paper and they were able to show generation of light at the double frequency. An amusing anecdote is that when they published their results in the prestigious journal *Physical Review Letters*, the editor mistook the small dot on the photographic paper at 347 nm for a spot of dirt and removed it from the article. Since then the field has grown continuously and second-harmonic generation has become a very powerful characterization tool with applications in biology, chemistry, and physics.

In the past a lot of interest in second-order nonlinear optical processes was motivated by their potential applications. Meanwhile, harmonic generation was found to be a very powerful characterization tool for the study of surfaces and interfaces. Indeed, due to symmetry reasons, second-harmonic generation is surface specific and has evolved into a mature characterization technique. Recently, sum-frequency generation also has matured into a reliable technique to probe surfaces and interfaces. However, whereas second-harmonic generation usually probes electronic transitions at interfaces, sum-frequency generation is typically used to address vibrational transitions. Both techniques are almost exclusively used by those with a background in nonlinear optics and remain largely unknown techniques for most scientists. With this book we would like to give the nonspecialist reader a taste of these exceptional characterization tools.

Several books on nonlinear optics have been published over the past 40 years. However, these books are often aimed at physicists and require a good understanding of optics, quantum theory, and nonlinear optical processes. Furthermore, several of these books focus on applications or materials design, instead of focusing on the potential of second-order nonlinear optics as a powerful characterization tool. However, many chemists, biochemists, biologists, and materials scientists are interested in using nonlinear optical techniques for characterization of their samples but often do not have the necessary background to become involved in this field. Also, for students at the PhD level who enter the field of nonlinear optics, we believe this book could be very helpful. Hence, there is obviously a need for a multidisciplinary book on this topic, without going too much into mathematical detail. This book is aimed toward a more general readership and provides an elementary description of nonlinear optics and several practical examples of how to use nonlinear optics as a versatile characterization tool. It provides a detailed description on how to implement nonlinear optical techniques for specific purposes, going from determination of molecular symmetries over surface characterization to biological imaging. The aim of the book is not to give a complete overview of all possible nonlinear optical techniques, but instead to focus on those techniques that are relatively easy to describe and implement in the lab.

Although it is usually preferred to use the International System of Units (SI), we have decided to define equations in the Gaussian unit system. The reason is that the majority of literature in the field of nonlinear optics still uses the Gaussian unit system. Furthermore, we believe that equations defined in the Gaussian unit system are physically more meaningful and straightforward. However, the reader probably easily understands SI units and therefore we will occasionally use SI units when discussing the optical parameters. Conversion factors from Gaussian to SI units can be found in any standard textbook on physics and optics, but for completeness we have included an appendix that briefly addresses this topic.

The outline of the book is as follows. Chapter 1 starts with a short introduction of linear optics from a perspective of polarizability and linear susceptibility. Both are fundamental parameters in the description of linear optical processes such as refraction and absorption. At the end of the chapter the linear optical formalism is extended by adding nonlinear terms that describe the second-order nonlinear optical process. Special attention is given to nonlinear susceptibility, a materials parameter that provides a framework for the description of second-order nonlinear optical phenomena.

Chapter 2 deals with incoherent second-harmonic generation, generated during nonlinear light scattering or hyper-Rayleigh scattering. In contrast to its linear analogue, scattered light is doubled in frequency. Since its discovery in the 1960s, little attention was paid to this very weak and "forbidden" process until the early 1990s when it was used as a simple and efficient technique to determine the hyperpolarizability of organic molecules. Since then the number of groups using this technique has grown tremendously. Here we would like to focus on the sensitivity of this technique to molecular symmetry. By analyzing the polarization state of the scattered light, it is possible to determine molecular symmetries and to study supramolecular aggregation.

Chapters 3 and 4 deal with the study of surfaces and interfaces, exploiting the intrinsic surface sensitivity of second-harmonic generation and sum-frequency generation. Chapter 3 starts with a very general introduction and introduces the theoretical framework to analyze second-harmonic generation from surfaces. Applications include the determination of surface symmetry and molecular orientation. Chapter 4 focuses on chiral surfaces and the specific effects—such as second-harmonic generation circular dichroism and second-harmonic generation optical rotatory dispersion—that can be observed in these systems. Using a conceptually simple framework, we gradually introduce the reader to the exceptional optical effects that can be observed in these systems.

The book concludes with Chapter 5, which discusses second-order imaging techniques. The chapter starts with a short introduction to microscopy in general, followed by an in-depth discussion of two-photon fluorescence (TPF) microscopy and second-harmonic generation microscopy.

The Authors

Thierry Verbiest obtained his PhD in physical chemistry in 1993 at the Catholic University of Leuven. After a postdoc at the IBM Almaden Research Center (San José, California), he returned to the University of Leuven and became a professor in the department of chemistry. He has performed research on surfaces, nonlinear optical techniques, and materials. Currently he is involved in magneto-optics and second-harmonic generation from chiral supramolecular materials. He is author or coauthor of more than 100 papers in the field of second-order nonlinear optics.

Koen Clays obtained his PhD in chemistry in 1989 at the Catholic University of Leuven. After a postdoc at Eastman-Kodak (Rochester, New York), he obtained a professor position at the Department of Chemistry, Catholic University of Leuven, Belgium. Recently he also became an adjunct professor at the Department of Physics and Astronomy, Washington State University, Pullman, Washington. His research interests are linear and nonlinear optics of molecular and nanostructured materials. Currently he is involved with research on photonic crystals and complex organic materials for second-order nonlinear optics.

Vincent Rodriguez received his PhD in physical chemistry from the University of Bordeaux, France, in 1989. This was followed by postdoctoral studies with Professor H. Schmid at the University of Geneva, Switzerland, before he moved to the Institut Laue-Langevin (European

Neutron Research Facilities) in Grenoble, France, as a physicist working on the D16 diffractometer (WANS, SANS). In 1993, he joined the group of Claude Sourisseau at the University of Bordeaux 1 in the Laboratory of Spectroscopies of Molecules and Crystals as an assistant professor in the Department of Chemistry. He is presently a full professor in physical chemistry at the Institute of Molecular Sciences. His primary research was in the field of solid-state materials, and in 1998 he entered the field of nonlinear optics. Over the years, he has contributed in the field of vibrational spectros
copies as well as nonlinear optics techniques such as second-harmonic generation, hyper-Raleigh scattering, and hyper-Raman scattering. His recent areas of interest concern novel photonic materials for nonlinear optics applications and imaging, molecular switches, and molecular and supramolecular chirality.

chapter one

General aspects of second-order nonlinear optics

In this chapter we will give a qualitative overview of the interaction of light with matter, which is necessary to understand the origin of non-linear optical effects. A rigorous mathematical treatment is beyond the scope of this book and can be found in many standard textbooks on optics (Stratton 1941; Böttcher 1973). Instead we will focus on the basic principles underlying linear and nonlinear optics, bearing in mind the background and interest of the readers for which this book was written.

1.1 Linear optical phenomena

1.1.1 Interaction of light with matter

Light, or electro-magnetic (EM) radiation in general, can be described by a time and space varying electric (E) and magnetic (B) field. For example, the electric field component of a monochromatic wave can be written as

$$E(r,t) = E_0(e^{ik.r - i\omega t} + cc) \tag{1.1}$$

with a time-dependent phase term, ωt, and a space-dependent phase term, $k.r$. ω is the pulsation (equal to $2\pi v$ with v the frequency) of the EM field, $k = 2\pi n/\lambda = n\omega/c$ (with λ the wavelength, c the speed of light in vacuum, and n the refractive index) is the wave vector that indicates the direction of light propagation (Figure 1.1), and cc denotes complex conjugate.

Since any material can be thought of as a collection of charged particles—electrons and positively charged cores—the oscillating electric field will interact with these particles. The positive ones will tend to move in the direction of the field, whereas the negative ones move toward the opposite direction of the field. The positively charged cores, however, have much greater mass than the electrons, and for high optical frequencies (in the ultraviolet and visible region of the spectrum), the motions of the electrons are more significant since they are instantaneous (approximately a few femtoseconds). In dielectric media, this charge separation will lead to induced dipole moments (μ) that oscillate with the same frequency as the

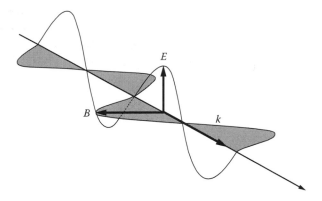

Figure 1.1 Schematic representation of an electro-magnetic wave.

applied optical field, and the bulk optical response is the resulting sum of the individual atomic or molecular responses. If the intensity of the incident light is sufficiently low, the relation between the induced dipole moment and the electric field $E(\omega)$ is given by

$$\mu(\omega) = \alpha(\omega)E(\omega) \tag{1.2}$$

with $\alpha(\omega)$ the molecular first-order polarizability or linear polarizability. The argument ω refers to the frequency of the electro-magnetic radiation. For clarity, this argument is often omitted, but the reader should always realize that the polarizability is a frequency-dependent quantity. Note also that the induced dipole moment will oscillate at the same frequency ω as the incident light field and as such acts as a source of radiation. The total dipole moment of the molecule (or molecular polarization) p is then the sum of the permanent dipole moment (μ_0) and the induced dipole moment ($\mu(\omega)$):

$$p = \mu_0 + \mu(\omega) \tag{1.3}$$

The induced polarization of the entire medium, arising from the sum of these induced dipole moments, is given by

$$P(\omega) = \sum_i \mu_i(\omega) = Nf_\omega \alpha(\omega)E(\omega) = \chi^{(1)}(\omega)E(\omega) \tag{1.4}$$

with $\chi^{(1)}(\omega)$ the first-order susceptibility (or linear susceptibility). N is the number density of molecules and f_ω a local field factor. The latter accounts for the effects of induced dipoles in the medium through electronic polarization. There are many different forms of local field factors but the most commonly used is the Lorentz–Lorenz correction factor given by

$$f_\omega = \frac{n_\omega^2 + 2}{3} \tag{1.5}$$

where n_ω is the refractive index at optical frequency ω. Note also that the total polarization of the medium is the sum of the spontaneous (or permanent) polarization $P^{(0)}$ and the induced polarization $P(\omega)$.

The first-order susceptibility contains all information about the optical properties of the macroscopic medium and describes processes such as dispersion, reflection, refraction, absorption, and scattering. As we will see, $\chi^{(1)}$ is, in general, a complex quantity $(\chi^{(1)} = \chi' + i\chi'')$ and related to the refractive index n_ω and the high frequency dielectric constant, ε_ω by

$$n_\omega^2 = \varepsilon_\omega = 1 + 4\pi\chi^{(1)}(\omega) \tag{1.6}$$

As a consequence, the refractive index is also a complex quantity, where the real part describes refraction and the imaginary part absorption of radiation.

1.1.2   Wave propagation in optical media

Any interaction of electromagnetic radiation with matter can be understood by using Maxwell's equations. In the Gaussian unit system these are given by

$$\nabla . D = 4\pi\rho$$

$$\nabla . B = 0$$

$$\nabla \times E = -\frac{1}{c}\frac{\partial B}{\partial t} \tag{1.7}$$

$$\nabla \times B = \frac{1}{c}\frac{\partial D}{\partial t} + \frac{4\pi}{c}J$$

where (E, B) is the field of the electromagnetic wave (E, electric field; B, magnetic induction field), D is the electric displacement, c is the speed of light in vacuum, ρ is the density of charges, and J is the density of current.

If we consider only materials that contain no free charge and no free current (dielectric medium), these equations reduce to

$$\nabla.D = 0$$

$$\nabla.B = 0$$

$$\nabla \times E = -\frac{1}{c}\frac{\partial B}{\partial t} \tag{1.8}$$

$$\nabla \times B = \frac{1}{c}\frac{\partial D}{\partial t}$$

The relation between the electric field (E) and electric displacement (D) is given by the first constitutive relation:

$$D = E + 4\pi P \tag{1.9}$$

Using Equation 1.4 this results in

$$D = E + 4\pi \chi^{(1)}E = (1 + 4\pi\chi^{(1)})E = n^2 E \tag{1.10}$$

which immediately shows that electric displacement and electric field are simply connected through the refractive index of the medium. Note that we have omitted the frequency argument for clarity. The relation (second constitutive relation) between magnetic induction field and magnetic field intensity is similarly given by

$$B = H + 4\pi M, \tag{1.11}$$

with M the magnetization of the medium. In nonmagnetic media this usually reduces to $B = H$.

To derive the optical wave equation, we proceed by taking the curl of $\nabla \times E = -\frac{1}{c}\frac{\partial B}{\partial t}$ and use $\nabla \times B = \frac{1}{c}\frac{\partial D}{\partial t}$ to obtain the general equation of propagation of the electric field $E(r,t)$:

$$\nabla \times (\nabla \times E(r,t)) + \frac{1}{c^2}\frac{\partial^2}{\partial t^2}D(r,t) = 0 \tag{1.12}$$

Note that we used the argument (r,t) to indicate the time (t) and spatial variation (r) of E and D. If we use the constitutive relation for D (Equation 1.9), this can be rewritten as

$$\nabla \times (\nabla \times E(r,t)) + \frac{1}{c^2}\frac{\partial^2}{\partial t^2}E(r,t) = -4\pi\frac{1}{c^2}\frac{\partial^2}{\partial t^2}P(r,t) \tag{1.13}$$

Hence, the polarization (P) is seen to act as a source term in the wave propagation equation. Substituting P with Equation 1.4 yields

$$\nabla \times (\nabla \times E(r,t)) + \frac{(1+4\pi\chi^{(1)})}{c^2} \frac{\partial^2}{\partial t^2} E(r,t) = \nabla \times (\nabla \times E(r,t)) + \frac{n^2}{c^2} \frac{\partial^2}{\partial t^2} E(r,t) = 0$$

(1.14)

Consequently, any propagating wave with velocity c/n and wave vector $k = n\frac{\omega}{c} = \frac{2\pi n \nu}{c}$ of the form $E(r,t) = E_0(e^{ik.r-i\omega t} + cc)$ will satisfy this equation. Hence, it is clear that the polarization of the material is the origin of the refractive index (n) of the material.

Let us generalize the meaning of the optical constant of a material by introducing its complex nature (Zernike and Midwinter 1973). In dielectric materials, the charged particles are bound together and the motion of the electrons varies in response to the electric field, $E(r,t)$, in a manner governed by the equation of motion for a classical harmonic oscillator. A simple description of the motion, $r(t)$, for one of the electrons is given in the following equation:

$$\frac{d^2r(t)}{dt^2} + 2\gamma\frac{dr(t)}{dt} + \omega_0^2 r(t) = -\frac{e}{m} E(r,t)$$

(1.15)

where ω_0 is the resonance frequency and γ is the damping constant. The term on the right-hand side of Equation 1.15 represents the force exerted on the electron by the applied field that drives the oscillation. If we take the applied optical field of the form

$$E(r,t) = E_0(e^{ik.r-i\omega t} + cc)$$

(1.16)

the solution to Equation 1.15 is of the form

$$r(t) = -\frac{e}{m} E_0 \frac{e^{ik.r-i\omega t}}{\omega_0^2 - 2i\gamma\omega - \omega^2} + cc$$

(1.17)

The motions of the collection of N electrons give rise to a macroscopic time-dependent polarization that reads

$$P(t) = -Ner(t)$$

(1.18)

and with Equation 1.4 we obtain the following expression for the linear susceptibility at frequency ω:

$$\chi^{(1)}(\omega) = \frac{Ne^2}{m} \frac{1}{(\omega_0^2 - 2i\gamma\omega - \omega^2)}$$

(1.19)

From this equation it is clear that $\chi^{(1)}$ is frequency dependent and is a complex quantity. The real and imaginary parts of this expression describe refraction and absorption, respectively:

$$\text{Re}[\chi^{(1)}(\omega)] = \chi'(\omega) = \frac{Ne^2}{m} \frac{\left(\omega_0^2 - \omega^2\right)}{\left(\omega^2 - \omega_0^2\right)^2 + (2\gamma\omega)^2} \tag{1.20}$$

$$\text{Im}[\chi^{(1)}(\omega)] = \chi''(\omega) = \frac{Ne^2}{m} \frac{2\gamma\omega}{\left(\omega^2 - \omega_0^2\right)^2 + (2\gamma\omega)^2} \tag{1.21}$$

The behavior of these expressions is illustrated in Figure 1.2 where real and imaginary parts of the susceptibility are plotted versus frequency in the region of an optical transition at ω_0. At ω_0, the absorption of χ'' is maximal and it has a Lorentzian shape since we consider an assembly of N identical oscillators. In that case, the spectral width of χ'' is called the homogeneous width. Going from low frequency to ω_0, the dispersion of χ' is connected to the increase of the backward dephasing of $P(\omega)$ with $E(\omega)$.

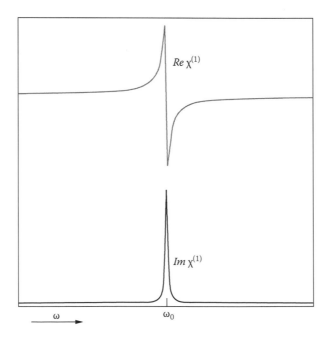

Figure 1.2 Variation of the real and imaginary parts of the linear susceptibility with frequency around a resonant frequency ω_0.

At resonance, a full π dephasing occurs that gives a forward delay for the propagation of the flux energy giving hence a local minimum in the dispersion curve of the real part of the susceptibility.

1.1.3 Tensor properties of polarizability and susceptibility

Up to now, we have treated polarizability and susceptibility as being simple scalar quantities. In general, however, they connect the different components of the electric field vector with the components of induced dipole moment or polarization, and as a consequence they should be considered as tensors of rank 2 with nine components $\chi_{ij}^{(1)}$. For example, the full expression for the polarization becomes:

$$P_x = \chi_{xx}^{(1)}E_x + \chi_{xy}^{(1)}E_y + \chi_{xz}^{(1)}E_z$$

$$P_y = \chi_{yx}^{(1)}E_x + \chi_{yy}^{(1)}E_y + \chi_{yz}^{(1)}E_z \tag{1.22}$$

$$P_z = \chi_{zx}^{(1)}E_x + \chi_{zy}^{(1)}E_y + \chi_{zz}^{(1)}E_z$$

or in matrix notation

$$\begin{bmatrix} P_x \\ P_y \\ P_y \end{bmatrix} = \begin{bmatrix} \chi_{xx}^{(1)} & \chi_{xy}^{(1)} & \chi_{xz}^{(1)} \\ \chi_{yx}^{(1)} & \chi_{yy}^{(1)} & \chi_{yz}^{(1)} \\ \chi_{zx}^{(1)} & \chi_{zy}^{(1)} & \chi_{zz}^{(1)} \end{bmatrix} \begin{bmatrix} E_x \\ E_y \\ E_z \end{bmatrix} \tag{1.23}$$

where we omitted the frequency arguments for clarity. A more convenient way of representing Equations 1.22 and 1.23 is

$$P_i = \sum_j \chi_{ij}^{(1)}E_j \tag{1.24}$$

The number of non-vanishing tensor components is, however, dependent on the symmetry of the molecule. For example, in an isotropic medium, where the x-, y-, and z-directions are equivalent, the off-axis components are zero, and Equation 1.23 reduces to

$$\begin{bmatrix} P_x \\ P_y \\ P_y \end{bmatrix} = \begin{bmatrix} \chi_{xx}^{(1)} & 0 & 0 \\ 0 & \chi_{yy}^{(1)} & 0 \\ 0 & 0 & \chi_{zz}^{(1)} \end{bmatrix} \begin{bmatrix} E_x \\ E_y \\ E_z \end{bmatrix} \tag{1.25}$$

with $\chi_{xx}^{(1)} = \chi_{yy}^{(1)} = \chi_{zz}^{(1)}$. Hence susceptibility (and polarizability) can be treated as a simple scalar quantity in isotropic media. In general, however, the number of non-vanishing components will be dependent on the symmetry of the medium, and a more elaborate discussion of the role of symmetry in optics will be given in Section 1.4.

1.2   Nonlinear optical phenomena

For very intense light (e.g., laser light), Equation 1.4 no longer holds, that is, the relation between polarization and electric field is no longer linear (Figure 1.3). In this nonlinear regime, the polarization is expanded in a Taylor series in terms of the total applied electric field. The induced polarization is then written as

$$P = P^{(1)} + P^{(2)} + P^{(3)} + \cdots = \chi^{(1)}E + \chi^{(2)}EE + \chi^{(3)}EEE + \cdots \tag{1.26}$$

where $P^{(1)}$ is linear in the electric field, $P^{(2)}$ is quadratic in the electric field, $P^{(3)}$ is cubic in the electric field, and so on.

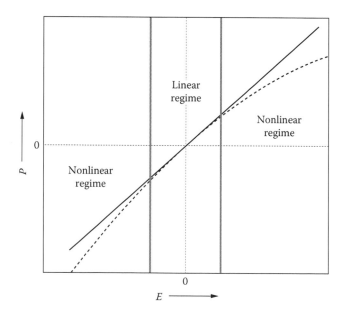

Figure 1.3 Linear and nonlinear responses of the electric polarization (P) with the strength of the input electric field (E).

$\chi^{(2)}$ and $\chi^{(3)}$ are the second- and third-order susceptibilities (or first- and second-nonlinear susceptibilities), characterizing the nonlinear optical response of the medium. Similarly, the induced dipole moment can be written as

$$\mu_{ind} = \mu^{(1)} + \mu^{(2)} + \mu^{(3)} + \cdots = \alpha E + \beta EE + \gamma EEE + \cdots \qquad (1.27)$$

with β and γ the second- and third-order polarizability (or first and second hyperpolarizability), respectively. Of special interest in this book are the quantities $\chi^{(2)}$ and β. Both quantities are responsible for second-order nonlinear optical phenomena that will be treated extensively in this book. For a system composed of N oscillators, the relation between them[*] is given by

$$\chi^{(2)} = Nf\beta \qquad (1.28)$$

where N is the number density and f a local field factor introduced previously.

Since susceptibility and polarizability connect the product of the two field vectors with the polarization vector or induced dipole moment, β and $\chi^{(2)}$ are tensors of rank 3. They contain 27 components and can be represented by a $3 \times 3 \times 3$ matrix. Later in this chapter we will deal with the symmetry properties of these specific tensors.

1.3 Examples of nonlinear optical phenomena

It can be easily seen that $\chi^{(2)}$ is responsible for nonlinear optical effects. If we focus on the term $P^{(2)}$ that is quadratically dependent on the electric field and combine it with the expression for the electric field $E(r, t) = E_0(e^{ik.r - i\omega t} + cc)$, we obtain:

$$P^{(2)} = \chi^{(2)} EE = \chi^{(2)} E_0(e^{ik.r - i\omega t} + cc) E_0(e^{ik.r - i\omega t} + cc) = 2\chi^{(2)} E_0^2 + \chi^{(2)} E_0^2 (e^{2ik.r - i2\omega t} + cc)$$

$$(1.29)$$

Hence, due to the nonlinear interaction, an additional frequency independent term and a term at the double frequency arise. The first one is a process called optical rectification (OR), and the second one is frequency doubling or second-harmonic generation (SHG). More generally, if we use two input fields at different frequencies, $E_1(r, t) =$

---

[*] This additivity rule involves no interacting effect.

$E_{0,1}(e^{ik_1.r-i\omega_1 t}+cc)$ and $E_2(r,t)=E_{0,2}(e^{ik_2.r-i\omega_2 t}+cc)$, the total incident field is $E(r,t)=E_{0,1}(e^{ik_1.r-i\omega_1 t}+cc)+E_{0,2}(e^{ik_2.r-i\omega_2 t}+cc)$ and we obtain:

$$P^{(2)} = \chi^{(2)}E_{0,1}^2(e^{i2k_1.r-i2\omega_1 t}+cc)+\chi^{(2)}E_{0,2}^2(e^{i2k_2.r-i2\omega_2 t}+cc)+2\chi^{(2)}E_{0,1}^2+2\chi^{(2)}E_{0,2}^{(2)}$$

$$+2\chi^{(2)}E_{0,1}E_{0,2}(e^{i(k_1+k_2).r-i(\omega_1+\omega_2)t}+cc)+2\chi^{(2)}E_{0,1}E_{0,2}(e^{i(k_1-k_2).r-i(\omega_1-\omega_2)t}+cc)$$

$$(1.30)$$

where we see—in addition to SHG and OR contributions—terms at $\omega_1 + \omega_2$ and $\omega_1 - \omega_2$ appearing, representing sum-frequency and difference-frequency generation (SFG and DFG), respectively. Written in terms of its different frequency Fourier components we obtain

$$P^{(2)} = P^{(2\omega_1)} + P^{(2\omega_2)} + P^{(0)} + P^{(\omega_1+\omega_2)} + P^{(\omega_1-\omega_2)} \qquad (1.31)$$

Important to note is the factor of 2 that appears in front of the sum- and difference-frequency terms. For that reason, the general expression for the nonlinear polarization is often written as $P^{(2)}(\omega_3) = K\chi^{(2)}E(\omega_1)E(\omega_2)$, where K is a degeneracy factor that is equal to 1 when both input fields have the same frequency and 2 for different frequencies.

A final second-order nonlinear optical effects is obtained when one of the interacting fields is a static electric field, $E(0)$. In that case

$$P^{(2)} = \chi^{(2)}E(0)E_0(e^{ik.r-i\omega t}+cc) \qquad (1.32)$$

representing the Pockels or linear electro-optic effect.

It is quite instructive to consider the polarization, including linear optical effects, and to define an effective susceptibility

$$P = \chi^{(1)}E + \chi^{(2)}EE = (\chi^{(1)}+\chi^{(2)}E)E = \chi_{eff}E \qquad (1.33)$$

If we now use Equation 1.6 for the refractive index of the medium, we immediately see that the nonlinear optical effects are a correction on the refractive index at very high optical fields and that the overall refractive index becomes dependent on the strength of the input field. For example, for the case of the linear electro-optic effect, this leads to

$$n^2 = 1+4\pi\chi_{eff} = 1+4\pi(\chi^{(1)}+\chi^{(2)}E(0)) \qquad (1.34)$$

where the index of refraction becomes dependent on the strength of the applied static field, $E(0)$.

1.4 Symmetries in second-order nonlinear optics

Hyperpolarizabilities and susceptibilities exhibit various types of symmetry that are of fundamental importance in nonlinear optics: permutation symmetry, time-reversal symmetry, and symmetry in space. The time-reversal and permutation symmetries are fundamental properties of the susceptibilities themselves, whereas the spatial symmetry of the susceptibility tensors reflects the structural properties of the nonlinear medium. Hence, for centrosymmetric materials, all tensor components $\chi_{ijk}^{(2)}$ are null. As a consequence, second-order nonlinear optical effects are not observed in centrosymmetric crystals. For a centrosymmetric system, the electric field and the polarization transform as $E \rightarrow -E$ and $P \rightarrow -P$ under the action of a center of inversion. Because of the principle of Neumann (1885) that states that a symmetry operation is required to leave the sign and magnitude of physical properties unchanged, $\chi^{(2)} \rightarrow \chi^{(2)}$. As a consequence, $P^{(2)} = \chi^{(2)}EE$ transforms into $-P^{(2)} = \chi^{(2)}(-E)(-E)$ and $\chi^{(2)}$ has to be zero.

In general, the number of independent components of $\chi^{(2)}$ that are nonzero is completely determined by symmetry. To clearly indicate the tensor properties of $\chi^{(2)}$, we will rewrite Equation 1.29 as

$$P_i^{(2)} = \sum_{j,k} \chi_{ijk}^{(2)} E_j E_k \tag{1.35}$$

where the subscripts ijk refer to Cartesian coordinates in the laboratory frame. $P_i^{(2)}$ is a component of the polarization vector, E_j are components of the electric field vector, and $\chi_{ijk}^{(2)}$ is a susceptibility tensor component. In the more general case, where the two excitations are nondegenerated (i.e., sum-frequency or difference-frequency processes), there are $3 \times 3 \times 3$ components and the second-order susceptibility tensor is preferentially described as a rectangular 9×3 matrix. Thus, Equation 1.35 explicitly reads

$$
\begin{bmatrix} P_x^{(2)} \\ P_y^{(2)} \\ P_z^{(2)} \end{bmatrix}
=
\begin{bmatrix}
\chi_{xxx}^{(2)} & \chi_{xyy}^{(2)} & \chi_{xzz}^{(2)} & \chi_{xyz}^{(2)} & \chi_{xzy}^{(2)} & \chi_{xzx}^{(2)} & \chi_{xxz}^{(2)} & \chi_{xxy}^{(2)} & \chi_{xyx}^{(2)} \\
\chi_{yxx}^{(2)} & \chi_{yyy}^{(2)} & \chi_{yzz}^{(2)} & \chi_{yyz}^{(2)} & \chi_{yzy}^{(2)} & \chi_{yzx}^{(2)} & \chi_{yxz}^{(2)} & \chi_{yxy}^{(2)} & \chi_{yyx}^{(2)} \\
\chi_{zxx}^{(2)} & \chi_{zyy}^{(2)} & \chi_{zzz}^{(2)} & \chi_{zyz}^{(2)} & \chi_{zzy}^{(2)} & \chi_{zzx}^{(2)} & \chi_{zxz}^{(2)} & \chi_{zxy}^{(2)} & \chi_{zyx}^{(2)}
\end{bmatrix}
\begin{bmatrix} E_x^2 \\ E_y^2 \\ E_z^2 \\ E_y E_z \\ E_z E_y \\ E_z E_x \\ E_x E_z \\ E_x E_y \\ E_y E_x \end{bmatrix}
\tag{1.36}
$$

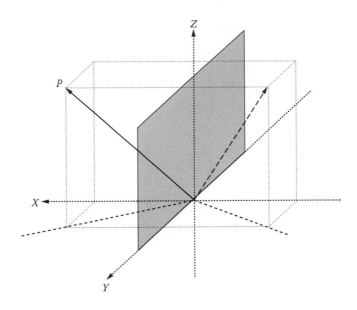

Figure 1.4 Effect of a mirror plane in the *y-z* plane on the electric polarization (*P*).

As an example, let us examine the influence of a mirror plane in the *y-z* plane on the second-order susceptibility. Under reflection in this plane (Figure 1.4), the coordinates transform as

$$x \rightarrow -x \qquad y \rightarrow y \qquad z \rightarrow z$$

Consequently, the polarization and field components transform as

$$E_x \rightarrow -E_x \qquad E_y \rightarrow E_y \qquad E_z \rightarrow E_z$$
$$P_x \rightarrow -P_x \qquad P_y \rightarrow P_y \qquad P_z \rightarrow P_z$$

The next step is to examine the effect of the mirror plane for each tensor component. For example, if we assume that all fields only have an *x*-component, the reflection will transform the relation $P_x = \chi_{xxx}^{(2)} E_x E_x$ into $-P_x = \chi_{xxx}^{(2)}(-E_x)(-E_x)$ or $P_x = -\chi_{xxx}^{(2)} E_x E_x$. Then, because of the principle of Neumann, $\chi_{xxx}^{(2)} = 0$. As a second example, consider the relation $P_x = \chi_{xxz}^{(2)} E_x E_z$. This will be transformed into $-P_x = \chi_{xxz}^{(2)}(-E_x)E_z$ or $P_x = \chi_{xxz}^{(2)} E_x E_z$. Therefore, since the sign of the tensor component remains unchanged, the component will not be canceled under reflection. The same procedure, considering all possible symmetry elements, has to be applied for the remaining 25 components.

To simplify this procedure, Fumi (1942) derived a technique called the "direct inspection" method. It was demonstrated that the components of a tensor transform as the product of their respective coordinates.

As a consequence, a tensor component will be zero if its index changes sign after a symmetry operation. In other words, the verification that $\chi_{xxx}^{(2)} = -\chi_{xxx}^{(2)}$ after the symmetry operation can be reduced to verifying that $xxx = -xxx$.

Hence, this procedure can be used to calculate all non-vanishing tensor components for each symmetry class (Table 1.1). Detailed indications about symmetry properties of these tensors may be found elsewhere (Butcher and Cotter 1990; Boyd 2003).

Additionally, in the special case of SHG, which is a degenerated case of SFG (the two excitation fields are indiscernible), a simple additional relation may be derived:

$$\chi_{ijk}^{(2)} = \chi_{ikj}^{(2)} \tag{1.37}$$

where *i*, *j*, and *k* are related to the axis orientations *x*, *y*, and *z*.

The more general symmetry requirement that is currently used for SHG, called the overall permutation symmetry, is an approximation that applies when all of the optical fields involved in the susceptibility formulae (excitations and response) are far removed from any transition. Kleinman (1962) first formulated it as:

$$\chi_{kji}^{(2)} = \chi_{ijk}^{(2)} \tag{1.38}$$

Finally, we would also like to note that for SHG, the tensor d_{ijk} is often used instead of the tensor $\chi_{ijk}^{(2)}$, the two tensors being equal.* Usually a "plane" representation of the Cartesian tensor d_{ijk} in the form d_{il} is used, where *i* = 1 corresponds to the *x* axis, *i* = 2 to the *y* axis, *i* = 3 to *z* axis, and *l* takes the following values:

	xx	yy	zz	yz = zy	xz = zx	xy = yx	
l =	1	2	3	4	5	6	(1.39)

Following this contracted notation, the polarization components can be written as:

$$\begin{bmatrix} P_x^{(2)} \\ P_y^{(2)} \\ P_z^{(2)} \end{bmatrix} = \begin{bmatrix} d_{11} & d_{12} & d_{13} & d_{14} & d_{15} & d_{16} \\ d_{21} & d_{22} & d_{23} & d_{24} & d_{25} & d_{26} \\ d_{31} & d_{32} & d_{33} & d_{34} & d_{35} & d_{36} \end{bmatrix} \begin{bmatrix} E_x^2 \\ E_y^2 \\ E_z^2 \\ 2E_yE_z \\ 2E_xE_z \\ 2E_xE_y \end{bmatrix} \tag{1.40}$$

---

* The relation $\chi^{(2)} = d$ is dependent on the choice of the convention. Here, we have adopted convention II from Kuzyk and Dirk (1998).

Table 1.1 Nonvanishing Tensor Components for Point Group Symmetries

International Notation	Schoenflies Notation	Nonvanishing Tensor Components Number of Independent Nonzero Elements
1	C_1	$\begin{bmatrix} xxx & xyy & xzz & xyz & xzy & xzx & xxz & xxy & xyx \\ yxx & yyy & yzz & yyz & yzy & yzx & yxz & yxy & yyx \\ zxx & zyy & zzz & zyz & zzy & zzx & zxz & zxy & zyx \end{bmatrix}$ (27)
$\bar{1}$	C_i	$\begin{bmatrix} 0 & 0 & 0 & 0 & 0 & 0 & 0 & 0 & 0 \\ 0 & 0 & 0 & 0 & 0 & 0 & 0 & 0 & 0 \\ 0 & 0 & 0 & 0 & 0 & 0 & 0 & 0 & 0 \end{bmatrix}$ (0)
m (xz plane)	C_s	$\begin{bmatrix} xxx & xyy & xzz & 0 & 0 & xzx & xxz & 0 & 0 \\ 0 & 0 & 0 & yyz & yzy & 0 & 0 & yxy & yyx \\ zxx & zyy & zzz & 0 & 0 & zzx & zxz & 0 & 0 \end{bmatrix}$ (14)
2 (y axis)	C_2	$\begin{bmatrix} 0 & 0 & 0 & xyz & xzy & 0 & 0 & xxy & xyx \\ yxx & yyy & yzz & 0 & 0 & yzx & yxz & 0 & 0 \\ 0 & 0 & 0 & zyz & zzy & 0 & 0 & zxy & zyx \end{bmatrix}$ (13)
2/m	C_{2h}	$\begin{bmatrix} 0 & 0 & 0 & 0 & 0 & 0 & 0 & 0 & 0 \\ 0 & 0 & 0 & 0 & 0 & 0 & 0 & 0 & 0 \\ 0 & 0 & 0 & 0 & 0 & 0 & 0 & 0 & 0 \end{bmatrix}$ (0)
mm2 (z axis)	C_{2v}	$\begin{bmatrix} 0 & 0 & 0 & 0 & 0 & xzx & xxz & 0 & 0 \\ 0 & 0 & 0 & yyz & yzy & 0 & 0 & 0 & 0 \\ zxx & zyy & zzz & 0 & 0 & 0 & 0 & 0 & 0 \end{bmatrix}$ (7)
222	D_2	$\begin{bmatrix} 0 & 0 & 0 & xyz & xzy & 0 & 0 & 0 & 0 \\ 0 & 0 & 0 & 0 & 0 & yzx & yxz & 0 & 0 \\ 0 & 0 & 0 & 0 & 0 & 0 & 0 & zxy & zyx \end{bmatrix}$ (6)
mmm	D_{2h}	$\begin{bmatrix} 0 & 0 & 0 & 0 & 0 & 0 & 0 & 0 & 0 \\ 0 & 0 & 0 & 0 & 0 & 0 & 0 & 0 & 0 \\ 0 & 0 & 0 & 0 & 0 & 0 & 0 & 0 & 0 \end{bmatrix}$ (0)
4	C_4	$\begin{bmatrix} 0 & 0 & 0 & xyz & xzy & xzx & xxz & 0 & 0 \\ 0 & 0 & 0 & xxz & xzx & yzx & yxz & 0 & 0 \\ zxx & zxx & zzz & 0 & 0 & 0 & 0 & zxy & zyx \end{bmatrix}$ (7)

(Continued)

Table 1.1 Nonvanishing Tensor Components for Point Group Symmetries
(Continued)

International Notation	Schoenflies Notation	Nonvanishing Tensor Components — Number of Independent Nonzero Elements	
$\bar{4}$	S_4	$\begin{bmatrix} 0 & 0 & 0 & xyz & xzy & xzx & xxz & 0 & 0 \\ 0 & 0 & 0 & \overline{xxz} & \overline{xzx} & xzy & xyz & 0 & 0 \\ zxx & \overline{zxx} & 0 & 0 & 0 & 0 & 0 & zxy & zyx \end{bmatrix}$	(6)
$4/m$	C_{4h}	$\begin{bmatrix} 0 & 0 & 0 & 0 & 0 & 0 & 0 & 0 & 0 \\ 0 & 0 & 0 & 0 & 0 & 0 & 0 & 0 & 0 \\ 0 & 0 & 0 & 0 & 0 & 0 & 0 & 0 & 0 \end{bmatrix}$	(0)
$4mm$	C_{4v}	$\begin{bmatrix} 0 & 0 & 0 & 0 & 0 & xzx & xxz & 0 & 0 \\ 0 & 0 & 0 & xxz & xzx & 0 & 0 & 0 & 0 \\ zxx & zxx & zzz & 0 & 0 & 0 & 0 & 0 & 0 \end{bmatrix}$	(4)
$\bar{4}2m$	D_{2d}	$\begin{bmatrix} 0 & 0 & 0 & xyz & xzy & 0 & 0 & 0 & 0 \\ 0 & 0 & 0 & 0 & 0 & xzy & xyz & 0 & 0 \\ 0 & 0 & 0 & 0 & 0 & 0 & 0 & zxy & zxy \end{bmatrix}$	(3)
422	D_4	$\begin{bmatrix} 0 & 0 & 0 & xyz & xzy & 0 & 0 & 0 & 0 \\ 0 & 0 & 0 & 0 & 0 & \overline{xzy} & \overline{xyz} & 0 & 0 \\ 0 & 0 & 0 & 0 & 0 & 0 & 0 & zxy & \overline{zxy} \end{bmatrix}$	(3)
$4/mmm$	D_{4h}	$\begin{bmatrix} 0 & 0 & 0 & 0 & 0 & 0 & 0 & 0 & 0 \\ 0 & 0 & 0 & 0 & 0 & 0 & 0 & 0 & 0 \\ 0 & 0 & 0 & 0 & 0 & 0 & 0 & 0 & 0 \end{bmatrix}$	(0)
3	C_3	$\begin{bmatrix} xxx & \overline{xxx} & 0 & xyz & xzy & xzx & xxz & \overline{yyy} & \overline{yyy} \\ \overline{yyy} & yyy & 0 & xxz & xzx & xzy & xyz & xxx & \overline{xxx} \\ zxx & zxx & zzz & 0 & 0 & 0 & 0 & zxy & \overline{zxy} \end{bmatrix}$	(9)
$\bar{3}$	S_6	$\begin{bmatrix} 0 & 0 & 0 & 0 & 0 & 0 & 0 & 0 & 0 \\ 0 & 0 & 0 & 0 & 0 & 0 & 0 & 0 & 0 \\ 0 & 0 & 0 & 0 & 0 & 0 & 0 & 0 & 0 \end{bmatrix}$	(0)
$3m$	C_{3v}	$\begin{bmatrix} 0 & 0 & 0 & 0 & 0 & xzx & xxz & \overline{yyy} & \overline{yyy} \\ \overline{yyy} & yyy & 0 & xxz & xzx & 0 & 0 & 0 & 0 \\ zxx & zxx & zzz & 0 & 0 & 0 & 0 & 0 & 0 \end{bmatrix}$	(5)

(*Continued*)

Table 1.1 Nonvanishing Tensor Components for Point Group Symmetries (Continued)

International Notation	Schoenflies Notation	Nonvanishing Tensor Components / Number of Independent Nonzero Elements
32	D_3	$\begin{bmatrix} xxx & \overline{xxx} & 0 & xyz & xzy & 0 & 0 & 0 & 0 \\ 0 & 0 & 0 & 0 & 0 & \overline{xzy} & \overline{xyz} & xxx & \overline{xxx} \\ 0 & 0 & 0 & 0 & 0 & 0 & 0 & zxy & \overline{zxy} \end{bmatrix}$ (4)
$\overline{3}m$	D_{3d}	$\begin{bmatrix} 0 & 0 & 0 & 0 & 0 & 0 & 0 & 0 & 0 \\ 0 & 0 & 0 & 0 & 0 & 0 & 0 & 0 & 0 \\ 0 & 0 & 0 & 0 & 0 & 0 & 0 & 0 & 0 \end{bmatrix}$ (0)
$\overline{6}$	C_{3h}	$\begin{bmatrix} xxx & \overline{xxx} & 0 & 0 & 0 & 0 & 0 & \overline{yyy} & \overline{yyy} \\ \overline{yyy} & yyy & 0 & 0 & 0 & 0 & 0 & xxx & xxx \\ 0 & 0 & 0 & 0 & 0 & 0 & 0 & 0 & 0 \end{bmatrix}$ (2)
6	C_6	$\begin{bmatrix} 0 & 0 & 0 & xyz & xzy & xzx & xxz & 0 & 0 \\ 0 & 0 & 0 & xxz & xzx & \overline{xzy} & \overline{xyz} & 0 & 0 \\ zxx & zxx & zzz & 0 & 0 & 0 & 0 & zxy & \overline{zxy} \end{bmatrix}$ (7)
$6/m$	C_{6h}	$\begin{bmatrix} 0 & 0 & 0 & 0 & 0 & 0 & 0 & 0 & 0 \\ 0 & 0 & 0 & 0 & 0 & 0 & 0 & 0 & 0 \\ 0 & 0 & 0 & 0 & 0 & 0 & 0 & 0 & 0 \end{bmatrix}$ (0)
$\overline{6}m2$	D_{3h}	$\begin{bmatrix} 0 & 0 & 0 & 0 & 0 & 0 & 0 & \overline{yyy} & \overline{yyy} \\ \overline{yyy} & yyy & 0 & 0 & 0 & 0 & 0 & 0 & 0 \\ 0 & 0 & 0 & 0 & 0 & 0 & 0 & 0 & 0 \end{bmatrix}$ (1)
$6mm$	C_{6v}	$\begin{bmatrix} 0 & 0 & 0 & 0 & 0 & xzx & xxz & 0 & 0 \\ 0 & 0 & 0 & xxz & xzx & 0 & 0 & 0 & 0 \\ zxx & zxx & zzz & 0 & 0 & 0 & 0 & 0 & 0 \end{bmatrix}$ (4)
622	D_6	$\begin{bmatrix} 0 & 0 & 0 & xyz & xzy & 0 & 0 & 0 & 0 \\ 0 & 0 & 0 & 0 & 0 & \overline{xzy} & \overline{xyz} & 0 & 0 \\ 0 & 0 & 0 & 0 & 0 & 0 & 0 & zxy & \overline{zxy} \end{bmatrix}$ (3)

(Continued)

Table 1.1 Nonvanishing Tensor Components for Point Group Symmetries
(Continued)

International Notation	Schoenflies Notation	Nonvanishing Tensor Components Number of Independent Nonzero Elements	
6/mmm	D₆ₕ	$\begin{bmatrix} 0 & 0 & 0 & 0 & 0 & 0 & 0 & 0 & 0 \\ 0 & 0 & 0 & 0 & 0 & 0 & 0 & 0 & 0 \\ 0 & 0 & 0 & 0 & 0 & 0 & 0 & 0 & 0 \end{bmatrix}$	(0)
23	T	$\begin{bmatrix} 0 & 0 & 0 & xyz & xzy & 0 & 0 & 0 & 0 \\ 0 & 0 & 0 & 0 & 0 & xyz & xzy & 0 & 0 \\ 0 & 0 & 0 & 0 & 0 & 0 & 0 & xyz & xzy \end{bmatrix}$	(2)
m3̄	Tₕ	$\begin{bmatrix} 0 & 0 & 0 & 0 & 0 & 0 & 0 & 0 & 0 \\ 0 & 0 & 0 & 0 & 0 & 0 & 0 & 0 & 0 \\ 0 & 0 & 0 & 0 & 0 & 0 & 0 & 0 & 0 \end{bmatrix}$	(0)
4̄3m	T_d	$\begin{bmatrix} 0 & 0 & 0 & xyz & xyz & 0 & 0 & 0 & 0 \\ 0 & 0 & 0 & 0 & 0 & xyz & xyz & 0 & 0 \\ 0 & 0 & 0 & 0 & 0 & 0 & 0 & xyz & xyz \end{bmatrix}$	(1)
432	O	$\begin{bmatrix} 0 & 0 & 0 & xyz & \overline{xyz} & 0 & 0 & 0 & 0 \\ 0 & 0 & 0 & 0 & 0 & xyz & \overline{xyz} & 0 & 0 \\ 0 & 0 & 0 & 0 & 0 & 0 & 0 & xyz & \overline{xyz} \end{bmatrix}$	(1)
m3m	Oₕ	$\begin{bmatrix} 0 & 0 & 0 & 0 & 0 & 0 & 0 & 0 & 0 \\ 0 & 0 & 0 & 0 & 0 & 0 & 0 & 0 & 0 \\ 0 & 0 & 0 & 0 & 0 & 0 & 0 & 0 & 0 \end{bmatrix}$	(0)
∞	C∞	$\begin{bmatrix} 0 & 0 & 0 & xyz & xzy & xzx & xxz & 0 & 0 \\ 0 & 0 & 0 & yyz & yzy & yzx & yxz & 0 & 0 \\ zxx & zyy & zzz & 0 & 0 & 0 & 0 & zxy & zyx \end{bmatrix}$	(7)
∞2	D∞	$\begin{bmatrix} 0 & 0 & 0 & xyz & xzy & 0 & 0 & 0 & 0 \\ 0 & 0 & 0 & 0 & 0 & \overline{xzy} & \overline{xyz} & 0 & 0 \\ 0 & 0 & 0 & 0 & 0 & 0 & 0 & zxy & \overline{zxy} \end{bmatrix}$	(3)
∞/m	C∞ₕ	$\begin{bmatrix} 0 & 0 & 0 & 0 & 0 & 0 & 0 & 0 & 0 \\ 0 & 0 & 0 & 0 & 0 & 0 & 0 & 0 & 0 \\ 0 & 0 & 0 & 0 & 0 & 0 & 0 & 0 & 0 \end{bmatrix}$	(0)

(*Continued*)

Table 1.1 Nonvanishing Tensor Components for Point Group Symmetries (Continued)

International Notation	Schoenflies Notation	Nonvanishing Tensor Components Number of Independent Nonzero Elements
∞mm	$C_{\infty v}$	$\begin{bmatrix} 0 & 0 & 0 & 0 & 0 & xzx & xxz & 0 & 0 \\ 0 & 0 & 0 & xxz & xzx & 0 & 0 & 0 & 0 \\ zxx & zxx & zzz & 0 & 0 & 0 & 0 & 0 & 0 \end{bmatrix}$ (4)
∞ / mm	$D_{\infty h}$	$\begin{bmatrix} 0 & 0 & 0 & xyz & xzy & xzx & xxz & 0 & 0 \\ 0 & 0 & 0 & \overline{xxz} & \overline{xzx} & xzy & xyz & 0 & 0 \\ zxx & \overline{zxx} & 0 & 0 & 0 & 0 & 0 & zxy & zyx \end{bmatrix}$ (6)

Note: Components *ijk* with negative sign are denoted by $\overline{ijk}$.

For SHG, only 18 independent elements exist and when Kleinman symmetry is valid, only 10 of them are independent.[*]

As an example, for a layer with symmetry 6 mm (C_{6v}) the form of the d_{ij} matrix (see Table 1.1) is

$$d_{ij} = \begin{pmatrix} 0 & 0 & 0 & 0 & d_{15} & 0 \\ 0 & 0 & 0 & d_{15} & 0 & 0 \\ d_{31} & d_{31} & d_{33} & 0 & 0 & 0 \end{pmatrix} \tag{1.41}$$

Note if Kleinman symmetry holds, then $d_{15} = d_{31}$ and we are left with two independent nonzero elements d_{31} and d_{33}. In this example, we clearly see that the higher the symmetry, the lower the number of independent parameters.

As can be seen from Table 1.1, all non-vanishing tensor components are given with their Cartesian components. As a consequence, the specific value of each component will depend on the choice of the coordinate system. Often, one chooses certain crystallographic or molecular axes as the coordinate axes and often this is sufficient to describe the properties of the material in a straightforward way. For systems with very low symmetry, however, the choice of coordinate system becomes more difficult and can lead to a lot of confusion. A convenient way to address this problem is to use the concept of rotational invariance. It allows for the treatment of the

---

[*] When Kleinman symmetry is valid we have (see Eq. 1.40): $d_{15} = d_{31}$, $d_{16} = d_{21}$, $d_{24} = d_{32}$, $d_{26} = d_{12}$, $d_{34} = d_{23}$, $d_{35} = d_{13}$, $d_{36} = d_{25} = d_{14}$.

nonlinear optical properties in terms of materials constants— irreducible tensors—that are independent of the coordinate system.

For example, a tensor of rank 3 can be written as a sum of irreducible representations (or irreducible tensors), each of rank 3, as (Coope et al. 1965):

$$T_3 = \sum_{\tau,J} {}^{\tau}T_3^J = T_3^0 + {}^1T_3^1 + {}^2T_3^1 + {}^3T_3^1 + {}^1T_3^2 + {}^2T_3^2 + T_3^3 \qquad (1.42)$$

This equation is the reduction spectrum of a third order rank tensor. The subscript indicates the rank of each irreducible tensor (3 in this case), the last superscript (J) gives the weight of each tensor, and the first superscript (τ) distinguishes between tensors of the same weight. It can be shown that each irreducible tensor of rank 3 and weight J can be expressed as an irreducible tensor of rank J. This is called the natural form of the tensor. For example, the natural form of T_3^0 is T_0^0 (rank 0). Natural tensors of rank (and weight) 0, 1, 2, and 3 are called scalars, vectors, deviators, and septors, respectively. The corresponding physical properties behave as multipoles of order 2^J. Multipoles of order 2^0, 2^1, 2^2, and 2^3 are called scalars, dipoles, quadrupoles, and octupoles, respectively. The reduction procedure of tensors of arbitrary rank has been thoroughly described in literature and an in-depth discussion is beyond the scope of this book. Further, the description of the hyperpolarizability and nonlinear susceptibility in terms of irreducible components can be quite involved. However, it is instructive to have a closer look at some specific examples.

If we consider a molecule with a C_{2v}-symmetry, for example, *p*-nitroaniline (Figure 1.5), there are, according to Table 1.1 and under the assumption of Kleinman symmetry, three independent non-vanishing tensor components—β_{zzz}, $\beta_{zxx} = \beta_{xxz} = \beta_{xzx}$, and $\beta_{zyy} = \beta_{yyz} = \beta_{yzy}$. Furthermore, the out-of-plane zyy-type components can be considered to be very small because of the limited polarizability in the y-direction. Therefore, we

Figure 1.5 *p*-Nitroaniline (PNA) and triaminotrinitrobenzene (TATB) in the molecular coordinate system xyz.

finally end up with two independent components β_{zzz} and $\beta_{zxx} = \beta_{xxz} = \beta_{xzx}$. For this particular system, the reduction spectrum of the hyperpolarizability tensor becomes

$$\beta = \beta_3 = {}^1\beta_3^1 + {}^2\beta_3^1 + {}^3\beta_3^1 + \beta_3^3 = 3\beta_3^1 + \beta_3^3 \tag{1.43}$$

with only irreducible tensors corresponding to a vector and a septor part. The vector part is given by

$$\beta_3^1 =$$

$$\frac{1}{15}\begin{bmatrix} 0 & 0 & 0 & 0 & 0 & \beta_{zzz}+\beta_{zxx} & \beta_{zzz}+\beta_{zxx} & 0 & 0 \\ 0 & 0 & 0 & \beta_{zzz}+\beta_{zxx} & \beta_{zzz}+\beta_{zxx} & 0 & 0 & 0 & 0 \\ \beta_{zzz}+\beta_{zxx} & \beta_{zzz}+\beta_{zxx} & 3(\beta_{zzz}+\beta_{zxx}) & 0 & 0 & 0 & 0 & 0 & 0 \end{bmatrix}$$

$$\tag{1.44}$$

In its natural form, this becomes

$$\beta_1^1 = \begin{bmatrix} 0 \\ 0 \\ \beta_{zzz}+\beta_{zxx} \end{bmatrix} \tag{1.45}$$

which represents the vectorial part of the hyperpolarizability with dipolar properties.

The septor part, with octupolar properties, is given by

$$\beta_3^3 =$$

$$\frac{1}{5}\begin{bmatrix} 0 & 0 & 0 & 0 & 0 & 4\beta_{zxx}-\beta_{zzz} & 4\beta_{zxx}-\beta_{zzz} & 0 & 0 \\ 0 & 0 & 0 & -\beta_{zzz}-\beta_{zxx} & -\beta_{zzz}-\beta_{zxx} & 0 & 0 & 0 & 0 \\ 4\beta_{zxx}-\beta_{zzz} & -\beta_{zzz}-\beta_{zxx} & 2\beta_{zzz}-3\beta_{zxx} & 0 & 0 & 0 & 0 & 0 & 0 \end{bmatrix}$$

$$\tag{1.46}$$

If we inspect Equation 1.44 it is clear that a molecule with $\beta_{zzz} = -\beta_{zxx} = -\beta_{xxz} = -\beta_{xzx}$ would lead to a vanishing vectorial part of the hyperpolarizability, but with a nonzero octupolar contribution. Such a molecule would not be compatible with dipolar properties, but would still exhibit a nonlinear optical response. Such a situation can be easily achieved by introducing a threefold axis in the molecule and the original C_{2v}-symmetry becomes a D_{3h}-symmetry. The resulting molecule is triaminotrinitrobenzene (TATB; Figure 1.5), a noncentrosymmetric molecule with only one independent hyperpolarizability component and no dipolar or vectorial properties (Zyss 1991). As a consequence, this type of molecule is usually referred to as octupolar molecules.

The careful reader will realize that the above is a very particular situation. In general, however, the presence of *iii*- and *ijj*-type components is indicative of dipolar or vectorial properties, both on a molecular scale and macroscopic scale. For example, the nonlinear optical response of molecules with a C_{2v} or $C_{\infty v}$ symmetry is typically dominated by such components and therefore the response is mainly dipolar. On a macroscopic scale, the SiO_2 α-quartz structure with D_3 symmetry is a typical example of polar structure where the dipolar *iii*-type component dominates the nonlinear optical response.

However, molecules or structures with a purely octupolar response are abundant. A typical example is CCl_4 with only *xyz*-type components. In that case, the reduction spectrum of the hyperpolarizability only contains a septor part. On a macroscopic scale, the crystal KH_2PO_4 with D_{2d} symmetry shows a purely octupolar response. In general, the *xyz*-type components are often referred to as octupolar components.

To summarize, from group theoretical consideration, it is possible to decompose third-rank tensors in a sum of so-called irreducible tensorial components referred to here as the dipolar and octupolar irreducible components (Maker 1970; Zyss 1993).

1.5   Second-order polarizabilities and susceptibilities

In dielectric materials with nonlinear optical response, the charged particles are bound together and the motion of the electrons varies in response to the electric field, $E(r,t)$, in a manner governed by the equation of motion for a classical anharmonic oscillator. A simple description of the motion, $r(t)$, for one of the electrons (but beyond the harmonic response described in Section 1.1 with Equation 1.15) is given in the following equation:

$$\frac{d^2r(t)}{dt^2}+2\gamma\frac{dr(t)}{dt}+\omega_0^2r+(\xi^{(2)}r^2(t)+\cdots)=-\frac{e}{m}E(r,t) \tag{1.47}$$

where ω_0 is the resonance frequency, γ is the damping constant, and $\xi^{(2)}$ is the first anharmonic term that characterizes the strength of the nonlinearity. The term on the right-hand side of Equation 1.47 represents the force exerted on the electron by the applied field that drives the oscillation. We assume that the applied optical field is of the form

$$E(r,t)=E_{0,1}(e^{ik_1.r-i\omega_1t}+cc)+E_{0,2}(e^{ik_2.r-i\omega_2t}+cc) \tag{1.48}$$

No general solution to Equation 1.47 for an applied electric field of the form Equation 1.48 is known. However, if the applied electric field is sufficiently weak, the nonlinear term $\xi^{(2)}r^2$ will be much smaller than the linear term $\omega_0^{(2)}r$ for any displacement $r(t)$ that can be induced by the field. Under this hypothesis, Equation 1.47 can be solved iteratively by means of a perturbation expansion that is extensively detailed elsewhere (Boyd 2003). Furthermore, the motions of the collection of N electrons give rise to a macroscopic time-dependent polarization of the form $P(t) = -Ner(t)$ that contains Fourier components at different frequencies. One of these Fourier components at frequency ω_3, $P^{(2)}(\omega_3)$, is responsible for SFG, where $\omega_3 = \omega_1 + \omega_2$ and $K = 2$ (or SHG with $\omega_1 = \omega_2$ and $K = 1$), from the sample and is proportional to the product of the two incident electric fields:

$$P^{(2)}(\omega_3) = K.\chi^{(2)}E(\omega_1)E(\omega_2) \tag{1.49}$$

with

$$\chi^{(2)}(\omega_3) = \xi^{(2)}\frac{Ne^3}{m^2}\frac{1}{\left(\omega_0^2 - 2i\gamma\omega_3 - \omega_3^2\right)\left(\omega_0^2 - 2i\gamma\omega_2 - \omega_2^2\right)\left(\omega_0^2 - 2i\gamma\omega_1 - \omega_1^2\right)} \tag{1.50}$$

The second-order susceptibility can also be expressed in terms of the product of linear susceptibilities (see Equation 1.19) as

$$\chi^{(2)}(\omega_3) = \xi^{(2)}\frac{m}{N^2e^3}\chi^{(1)}(\omega_3)\chi^{(1)}(\omega_2)\chi^{(1)}(\omega_1) \tag{1.51}$$

Miller (1964) noticed that the quantity

$$\frac{\chi^{(2)}(\omega_3)}{\chi^{(1)}(\omega_3)\chi^{(1)}(\omega_2)\chi^{(1)}(\omega_1)} \tag{1.52}$$

is nearly constant for all noncentrosymmetric crystals. Hence, the empirical rule of Miller can be understood here in terms of Equation 1.51, which shows that in a first approximation, the quantity $\xi^{(2)}\frac{m}{N^2e^3}$ may be a supposed constant in noncentrosymmetric crystals. Assuming a mean constant electronic number density N, approximately 10^{28} m^3, as well as similar nonlinear anharmonic restoring forces for any crystals that estimates $m\omega_0^2r \approx m\xi^{(2)}r^2$, an order of magnitude of the strength of the nonlinearity is given by

$$\xi^{(2)} \approx \frac{\omega_0^2}{r} \geq \frac{(10^{16})^2}{10^{-10}} \geq 10^{42}\,N\,Kg^{-1}m^{-2} \tag{1.53}$$

Since ω_0 and the displacement of the electron from its equilibrium position r (which is less than the interatomic distance $d \sim 10^{-10}$ m) are roughly the same in solids, the anharmonic constant $\xi^{(2)}$ would be expected to be the same for second-order active materials.

Therefore an estimate of the static (at zero frequency) second-order susceptibility can be obtained

$$\chi^{(2)}(0) = \zeta^{(2)} \frac{Ne^3}{m^2} \frac{1}{\omega_0^6} \approx \frac{Ne^3}{m^2 \omega_0^4 r} \approx pm/V \qquad (1.54)$$

The second-order susceptibility presents three important properties:

1. $\chi^{(2)}$ is proportional to the quadratic anharmonic constant $\xi^{(2)}$ whose value reflects the symmetry of the electronic density profile of the media. In isotropic material, the potential becomes symmetric and no second-order response occurs (in the electric-dipole approximation).

2. $\chi^{(2)}$ has resonances at frequencies $\omega_1 = \omega_0$, $\omega_2 = \omega_0$, and $\omega_3 = \omega_0$ for SFG. In the case of SHG, there are two resonances frequencies at ω_0 and $\omega_0/2$. Dispersion effects of the susceptibility are expected as well as enhanced responses close to resonance (Figure 1.6). The response of a specific medium may be selectively amplified by resonance or preresonance effects.

3. $\chi^{(2)}$ is proportional to the density number of electrons N. Therefore, second-order responses will be sensitive to the concentration of molecules and also to a modification of the electronic density of a surface or layer.

These simple considerations underline the ability of second-order experiments with molecular systems (bounded electrons) or even metals (less bounded electrons). The absence of the center of symmetry that contributes to the electric-dipolar susceptibility $\chi^{(2)}$ can be achieved in several ways. The conventional approach considers the macroscopic polarization within the material, which can be generated by noncentrosymmetric molecules. A macroscopic polarization results when dipolar molecules are oriented in the same direction, or when ions are organized in a noncentrosymmetric crystal structure. Examples of such materials are dielectric crystals and dipolar molecules aligned in an electric field. It is important to note that it is not sufficient to develop a material composed of dipolar entities since these components must be organized in a noncentrosymmetric fashion as well. In Chapter 3 we will give some worked out examples of such macroscopic polar assemblies that develop a quadratic nonlinear optical response.

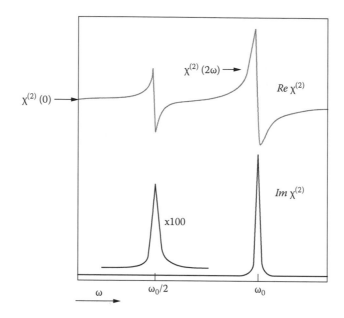

Figure 1.6 Variation of the real and imaginary parts of the second-order suscep-tibility in the SHG case with harmonic frequency around, 2ω, close to resonant frequency ω_0.

1.6   Beyond the electric-dipole approximation

The nonlinear interaction between the material and the electro-magnetic field can be treated to various degrees of detail. Up to now, we only con-sidered electric-dipole contributions to the nonlinear optical response, but, in general, higher-order contributions may occur. In this case, the second-order nonlinear polarization can be written as

$$P_i^{(2)} = \chi_{ijk}^{(2),eee}E_iE_j + \chi_{ijk}^{(2),eem}E_jB_k + \chi_{ijkl}^{(2),eeq}E_j\nabla_kE_l \tag{1.55}$$

where summation over repeated indices is implied and the superscripts associate the respective subscripts with electric-dipole (e), magnetic-dipole (m), and electric-quadrupole (q) interactions (note that two subscripts are associated with the quadrupole interaction). In addition, the material also develops a nonlinear magnetization

$$M^{(2)} = \chi_{ijk}^{mee}E_jE_k \tag{1.56}$$

and quadrupolarization

$$Q_{ij}^{(2)} = \chi_{ijkl}^{qee}E_kE_l \tag{1.57}$$

The nonlinear polarization, magnetization, and quadrupolarization all act as sources of radiation. Electric-dipole contributions are often referred to as local contributions, whereas magnetic-dipole and electric-quadrupole contributions are nonlocal. Note also that the tensors χ^{eem} and χ^{mee} are third-rank tensors, but with different symmetry properties than the tensor χ^{eee}. This is due to the fact that two indices of the *eem* (or *mee*) tensors are associated with a polar vector (the electric field), while the remaining coefficient is associated with an axial vector (magnetic field). Consequently, any improper transformation (reflection or inversion) leads to a different sign between the respective components of the *eee* tensor and the *mee* and *eem* tensors. The tensors *qee* and *eeq* are fourth-rank tensors for which all indices are associated with polar vectors. Note that both magnetic-dipole and electric-quadrupole contribution can also occur in centrosymmetric media due to their different symmetry properties. As a consequence, contributions from the bulk can also contribute to the total nonlinear optical response. This is, for example, the case in Silicon, where a significant portion of the second-harmonic signal is generated in the bulk. On the other hand, for most samples an electric-dipole contribution that arises from the surface will dominate the second-harmonic response.

In the literature, the total nonlinear optical response is often described by an effective polarization that includes magnetization and quadrupolarization:

$$P_{eff}^{(2)} = P^{(2)} + i\frac{c}{2\omega}\nabla \times M^{(2)} - \nabla.Q^{(2)} \tag{1.58}$$

Furthermore, in the literature, *qee* and *eeq* terms are often embedded in *eem* and *mee* terms. Thus, it is important to point out that these "new" *eem* and *mee* terms are in fact composite response tensor components. Nonlocal contributions have been widely discussed in the literature (Adler 1964; Shen 1984; Hoshi et al. 1995; Munn 1996). When electric dipole moments are induced by electric field gradients (or magnetic inductions), the electric quadrupole moments (or magnetic moments) are also induced by electric fields. Therefore, all the nonlinear electric polarization contributions produced at least by the magnetic dipole coupling (*eem* terms) and the electric quadrupole (*eeq* terms) may be expressed as an unique effective *eem* (or *eeq*) tensor. Additionally, the nonlinear magnetization contributions produced by the magnetic dipole (*mee* terms) and the electric quadrupole response (*qee* terms) may be expressed as a unique effective mee (or *qee*) tensor. All these points will be illustrated in Chapter 4 where effective *eem* and *mee* terms will be used.

References

Adler, E., *Phys. Rev.*, A3 (1964), A728.

Böttcher, C.J.F. *Theory of Electric Polarization*, 2nd ed., Amsterdam: Elsevier, 1973.

Boyd, R.W. *Nonlinear Optics*, 2nd ed., San Diego, CA: Academic Press, 2003.

Butcher, P.N., and D. Cotter, *The Elements of Nonlinear Optics* (Cambridge Studies in Modern Optics, vol. 9), Cambridge, UK: Cambridge University Press, 1990.

Coope, J.A.R., R.F. Snider, and F.R. McCourt, *J. Chem. Phys.*, 43 (1965), 2269.

Fumi, F.G., *Acta Cryst.*, 5 (1942), 44.

Hoshi, H., T. Yamada, K. Ishikawa, H. Takezoe, and A. Fukuda, *Phys. Rev.*, B52 (1995), 12355.

Kleinman, D.A., *Phys. Rev.*, 126 (1962), 1977.

Kuzyk, M.G., and C.W. Dirk, eds., *Characterization Techniques and Tabulations for Organic Nonlinear Optical Materials*, New York: Marcel Dekker, 1998.

Maker, P.D., *Phys. Rev. A*, 1 (1970), 923.

Miller, R.C., *Appl. Phys. Lett.*, 5 (1964): 17.

Munn, R.W., *Mol. Phys.*, 89 (1996), 555.

Neumann, F.E., *Vorlesungen über die Theorie der Elastizität der festen Körper und des Lichtäthers*, edited by O.E. Meyer, Leipzig: B. G. Teubner-Verlag, 1885.

Shen, Y.R., *The Principles of Nonlinear Optics*, New York: Wiley & Sons, 1984.

Stratton, J.A., *Electromagnetic Theory*, New York: McGraw-Hill, 1941.

Zernike, F., and J.E. Midwinter, *Applied Nonlinear Optics*, New York: Wiley, 1973.

Zyss, J., *Nonlinear Opt.*, 1 (1991), 3.

Zyss, J., *J. Chem. Phys.*, 98 (1993), 6583.

chapter two

Determination of molecular symmetry with hyper-Rayleigh scattering

2.1 Hyper-Rayleigh scattering: General principles

Hyper-Rayleigh scattering (HRS), also called harmonic light scattering (HLS), is the nonlinear analogue of the well-known Rayleigh scattering (RS; Figure 2.1). Whereas Rayleigh scattering involves only linear interaction, resulting in scattering at an identical frequency as the impinging radiation, for the description of hyper-Rayleigh scattering it is necessary to allow for the generation of new frequency components from the original frequencies on the incident radiation. Just as for any (linear or nonlinear) interaction of electromagnetic radiation with matter, this can be understood by using Maxwell's equations. HRS can be considered as one of the most important experimental techniques in molecular second-order nonlinear optics. It can be used to determine molecular nonlinearities, but, probably more important, it has proven to be a very powerful technique to determine molecular and supramolecular symmetries. Another technique that has been used frequently to determine the hyperpolarizability of molecules is electric field induced second-harmonic generation (EFISH). In this technique, dipolar molecules in solution are oriented in a static electric field to create the necessary macroscopic noncentrosymmetry. However, unlike HRS it is not possible to determine molecular symmetries. Furthermore, octupolar and ionic molecules cannot be measured by EFISH since orientation in an electric field is not possible for these systems.

Although "hyper-Rayleigh" scattering does not refer to any specific higher-order scattering, it is usually reserved for the scattering at the second-harmonic frequency (Clays and Persoons 1991). Since there is no static field involved, hyper-Rayleigh scattering is a second-order nonlinear optical effect, governed by the second-order nonlinear susceptibility. As for any even-order phenomenon, there is the noncentrosymmetric requirement. This means that no second-harmonic generation can come from a centrosymmetric medium. Only noncentrosymmetric media (dipolar or

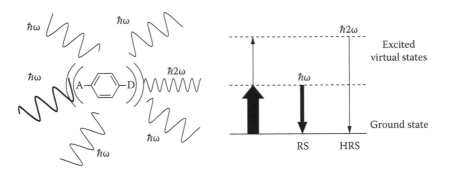

Figure 2.1 Rayleigh scattering (RS) and hyper-Rayleigh scattering (HRS) schemes.

octupolar) will give a second-order response. We should keep in mind that this rule is applied on the microscopic, molecular scale as well as on the macroscopic, bulk scale. At the molecular scale this problem can be solved by using electron donor and acceptor substituted conjugated D-π-A dipolar or octupolar molecules (CCl_4 for example; Terhune et al. 1965). However, when all such molecules are also randomly oriented, no second-order signal will occur from a bulk isotropic medium. A polar order should be induced on the macroscopic scale. This can be achieved by, for example, electric-field poling, noncentrosymmetric crystal growth, the X- or Z-type Langmuir–Blodgett deposition technique, or the more stable Y-type deposition but then in combination with alternate deposition (Prasad and Williams 1991), again to overcome the centrosymmetry arrangement (see Chapter 3).

Hyper-Rayleigh scattering does not appear to obey Neumann's principle (spatial symmetry requirement), as a second-order signal is generated from a simple solution, that is, an isotropic ensemble of solute molecules randomly dissolved in a solvent. However, it is only on average time and space that the solution is centrosymmetric. Due to density and rotational (particularly in liquids) fluctuations, locally in time and space, there are deviations away from centrosymmetry, allowing a (weak) second-order signal to appear (called incoherent response), in contrast to a bulk polar arrangement that gives a strong SHG signal (called coherent response).[*] Simply speaking, in a light scattering experiment (e.g., HRS) the time average response during the optical excitation (from a ground state to a virtual excited state) is sufficiently short (a few femtoseconds with a visible/near

---

[*] The notion of a coherent or incoherent signal arises from whether phase relations exist between the amplitudes of the fields radiated by an assembly of molecules.

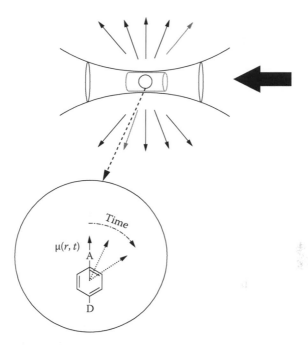

Figure 2.2 Hyper-Rayleigh scattering from a dipolar molecule. The time average of the fluctuating induced dipole response $\mu(r, t)$ during the excitation does not average to zero. At longer times the net dipole response is zero as it is expected in an isotropic medium.

infrared excitation) not to average to zero the net response of the induced dipole $\mu(r, t)$ within the radiating volume (typically a few cubic micrometers; Figure 2.2).* Thus, a dipolar molecule will give on average a HRS signal, whereas a centrosymmetric molecule (benzene with class symmetry D_{6h}, for example) will give no HRS signal.

The hyper-Rayleigh signal originates from the individual nonlinear scatterers that do not exhibit specific phase relations with respect to one another in the solution. Therefore, the signal is primarily related to the individual molecule's hyperpolarizability that is resulting in the molecules scattering field. Since there are no phase relations between the fields of the different molecules, we can arrive at the individual molecule's scattering intensity first by taking the square, and then at the total intensity by

---

* Strictly speaking, in the dipolar response, the scattered intensity is proportional to the dipolar autocorrelation response: $I_{HRS} \propto <\mu(r,0)\mu(r,t)>$, where the bracket indicates time average over the event.

multiplying with the number of molecules and applying the appropriate averaging over all possible orientations in the isotropic solution.

Note that in the case of specific phase relations between the scattering centers, one first has to take the sum of the fields (according to these phase relations) and only after arriving at the total varying electric field is one allowed to take the square to arrive at the intensity. For uncorrelated scatterers, the hyper-Rayleigh intensity is given by (Clays et al. 1994):

$$I_{2\omega} = \frac{16\,\pi^5}{c\,\lambda^4\,r^2}\,N\,f_\omega^4\,f_{2\omega}^2\,\langle \beta_{HRS}^2 \rangle\,I_\omega^2 \qquad (2.1)$$

where N is the concentration of chromophores, and f_ω and $f_{2\omega}$ are local field factors, as defined in Chapter 1. The brackets indicate the orientational averaging, λ the fundamental wavelength, r the distance to the scattering center, and c the speed of light in vacuum.

Note, however, that there is experimental evidence that some coherent contributions exist in the HRS signal. This is, for example, the case for neat CCl_4 (a structured liquid close to a rotator glass phase) in which a strong orientational correlation exists between neighboring molecules (Kaatz and Shelton 1996). As a result, 60% of the HRS signal comes from these effects. However, correlations are not so pronounced in most liquids. Recall that CCl_4 has an octupolar incoherent (monomolecular) HRS response, whereas most of the molecules that are investigated concern dipolar moieties. In the case of "classical" dipolar molecules, the dipolar incoherent response dominates the other contributions (coherent and so on) by circa 3 orders of magnitude. Finally in a binary solution, where the HRS response of chromophores is investigated, the response of diluted chromophores either dipolar (push-pull systems) or octupolar strongly dominate the net HRS response even if the solvent has an HRS signal. It has been found, however, that for high concentrations of polar molecules in apolar solvents, dimer formation results in centrosymmetric aggregates. This shows up as an HRS intensity that diminishes for higher chromophore concentrations.

The relationship between $\langle \beta_{HRS}^2 \rangle$ and the molecular tensor components β_{ijk} depends on the polarization state of both fundamental and harmonic light and also the scattering geometry. In classical HRS experiments, the 90° angle geometry is used exclusively. The fundamental light beam is propagating in the X-direction and polarized in the Z-direction, and the scattered light is collected in the Y-direction (Figure 2.3). Note that we distinguish between the laboratory coordinate system of reference (X, Y, Z), and the molecular coordinate system of reference (x, y, z).

In such a measuring geometry, the relationships (Figure 2.4) between the orientationally averaged tensor components (defined in the laboratory

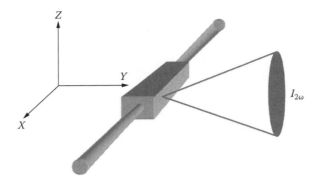

Figure 2.3 Schematic view of the classical 90° angle HRS geometry. An intense infrared laser beam is brought to focus in a cell containing the isotropic solution and the frequency-doubled light $I_{2\omega}$ is collected and detected under 90°.

frame XYZ) and the molecular tensor components (defined in the molecular internal frame xyz) are expressed as follows (Cyvin et al. 1965):

$$\langle \beta_{ZZZ}^2 \rangle = \frac{1}{7}\sum_i \beta_{iii}^2 + \frac{6}{35}\sum_{i \ne j} \beta_{iii}\beta_{ijj} + \frac{9}{35}\sum_{i \ne j} \beta_{ijj}^2 + \frac{6}{35}\sum_{ijk,cyclic} \beta_{iij}\beta_{jkk} + \frac{12}{35}\beta_{ijk}^2 \qquad (2.2)$$

$$\langle \beta_{XZZ}^2 \rangle = \frac{1}{35}\sum_i \beta_{iii}^2 - \frac{2}{105}\sum_{i \ne j} \beta_{iii}\beta_{ijj} + \frac{11}{105}\sum_{i \ne j} \beta_{ijj}^2 - \frac{2}{105}\sum_{ijk,cyclic} \beta_{iij}\beta_{jkk} + \frac{8}{35}\beta_{ijk}^2$$

$$(2.3)$$

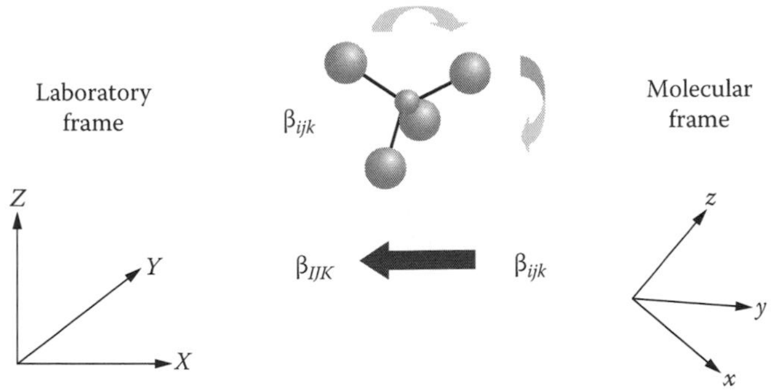

Isotropic averaging

Laboratory frame β_{ijk} Molecular frame

Z z

Y

β_{IJK} ⬅ β_{ijk} y

X x

Figure 2.4 Isotropic averaging of the molecular components of the hyperpolarizability.

The first subscript (X or Z) refers to the polarization state of the frequency-doubled light (in the laboratory coordinate system) and *ijk* refers to the molecular coordinate system. For an unpolarized scattered signal, since both polarizations are detected with equal sensitivity, the orientational average over β is the sum of both previous equations and

$$\langle \beta^2_{HRS} \rangle = \langle \beta^2_{ZZZ} \rangle + \langle \beta^2_{XZZ} \rangle \tag{2.4}$$

The orientational averaged hyperpolarizability squared $\langle \beta^2_{HRS} \rangle$ is related to the molecular hyperpolarizability tensor components according to Equations 2.2 and 2.3. For a polar molecule with C_{2v} symmetry (see, for example, PNA given in Chapter 1, Figure 1.5) these equations reduce to

$$\langle \beta^2_{ZZZ} \rangle = \frac{1}{7} \beta^2_{zzz} + \frac{6}{35} \beta_{zzz} \beta_{zxx} + \frac{9}{35} \beta^2_{zxx} \tag{2.5}$$

$$\langle \beta^2_{XZZ} \rangle = \frac{1}{35} \beta^2_{zzz} - \frac{2}{105} \beta_{zzz} \beta_{zxx} + \frac{11}{105} \beta^2_{zxx} \tag{2.6}$$

$$\langle \beta^2_{HRS} \rangle = \frac{6}{35} \beta^2_{zzz} + \frac{16}{105} \beta_{zzz} \beta_{zxx} + \frac{38}{105} \beta^2_{zxx} \tag{2.7}$$

This reduces to

$$\langle \beta^2_{HRS} \rangle \approx \left(\frac{1}{7} + \frac{1}{35} \right) \beta^2_{zzz} = \frac{6}{35} \beta^2_{zzz} \tag{2.8}$$

under the assumption that β_{zzz} is much larger than β_{zxx}. This is an assumption that is valid for most uniaxial dipolar molecules with the z-axis being the molecular dipole axis. Also from the way the HRS signal is dependent on the different hyperpolarizability tensor components and in combination with a typical 10% statistical error on the obtained values, it should be clear that the largest tensor element contributes by far the strongest to the signal. If the off-diagonal element is 10% of the largest diagonal element, then from Equation 2.7 it is clear that only 9% of the signal is caused by the contribution originating from this off-diagonal element.

For an octupolar molecule with D_{3h} symmetry (see, for example, TATB given in Chapter 1, Figure 1.5) only four equal tensor components remain. Indeed, as shown in Chapter 1, the hyperpolarizability tensor components in such system are $\beta_{zzz} = -\beta_{zxx} = -\beta_{xxz} = -\beta_{xzx}$ and we find:

$$\left\langle \beta_{ZZZ}^2 \right\rangle = \frac{1}{7}\beta_{zzz}^2 + \frac{6}{35}\beta_{zzz}\beta_{zxx} + \frac{9}{35}\beta_{xxz}^2 = \frac{8}{35}\beta_{zzz}^2 \tag{2.9}$$

$$\left\langle \beta_{XZZ}^2 \right\rangle = \frac{1}{35}\beta_{zzz}^2 - \frac{2}{105}\beta_{zzz}\beta_{zxx} + \frac{11}{105}\beta_{xxz}^2 = \frac{16}{105}\beta_{zzz}^2 \tag{2.10}$$

and

$$\left\langle \beta_{HRS}^2 \right\rangle = \frac{40}{105}\beta_{zzz}^2 \tag{2.11}$$

Only one tensor component β_{xyz} remains for an octupolar molecule with T_d symmetry, and we obtain:

$$\left\langle \beta_{ZZZ}^2 \right\rangle = \frac{12}{35}\beta_{xyz}^2 \tag{2.12}$$

$$\left\langle \beta_{XZZ}^2 \right\rangle = \frac{8}{35}\beta_{xyz}^2 \tag{2.13}$$

and therefore:

$$\left\langle \beta_{HRS}^2 \right\rangle = \frac{20}{35}\beta_{xyz}^2 \tag{2.14}$$

In anticipation of Section 2.3, "Determination of molecular symmetries," here is the appropriate place to point to the experimental possibility, but also the necessity to probe certain aspects of the molecular symmetry to arrive at the right interpretation of the HRS signal in terms of the correct tensor elements. When looking at Equations 2.2 and 2.3, it should be clear that the depolarization ratio (ρ), defined as

$$\rho = \frac{\left\langle \beta_{XZZ}^2 \right\rangle}{\left\langle \beta_{ZZZ}^2 \right\rangle} \tag{2.15}$$

is different for different occurrences of nonzero tensor components (Heesink et al. 1993; Morrison et al. 1996). As a case in point, for the prototypical one-dimensional elongated donor-acceptor molecule with the single major diagonal tensor component β_{zzz}, the depolarization ratio (ρ) is identical to

$$\rho = \frac{\frac{1}{35}\beta^2_{zzz}}{\frac{1}{7}\beta^2_{zzz}} = \frac{1}{5} \qquad (2.16)$$

whereas for octupolar molecules, this ratio is always equal to 2/3, this being the case for D_{3h} symmetry with four nonzero but identical tensor components:

$$\rho = \frac{\frac{16}{105}\beta^2_{zzz}}{\frac{24}{105}\beta^2_{zzz}} = \frac{2}{3} \qquad (2.17)$$

just as well as for with T_d symmetry with only one tensor component β_{xyz} :

$$\rho = \frac{\frac{8}{35}\beta^2_{xyz}}{\frac{12}{35}\beta^2_{xyz}} = \frac{2}{3} \qquad (2.18)$$

This already shows one of the potentials of HRS for structure determination. Other possibilities will be elaborated in Section 2.3 but it was necessary to point here to this depolarization ratio, since it is only the combination of an HRS intensity with a depolarization measurement that allows for the analysis toward the value for a molecular hyperpolarizability tensor component. Although, very often, the effective approximate solution structure, dipolar or octupolar, can be estimated from the molecular formula and a correct a priori assumption can be made.

2.2   *Experimental techniques and equipment*

Since the second-order nonlinear optical effects from an average time and space centrosymmetrical isotropic solution are forbidden in the dipole approximation, it is not a photon-efficient technique. It is only based on the fluctuations away from centrosymmetry, locally in time and space, that second-order effects can be observed from solutions. Therefore, pulsed lasers and tight focusing are needed to arrive at the peak intensities that are needed to observe HRS in solution.

The prototypical pulsed laser that is being used for HRS is the low duty cycle Q-switched Neodymium YAG laser providing millijoule energy in nanosecond pulses, typically resulting in megawatt peak power for subwatt average power (Clays and Persoons 1992). This laser is also providing a convenient near-infrared wavelength (1064 nm), resulting in a second-harmonic wavelength in a sensitive spectral region for photon detection by efficient photomultiplier tubes. The original way of shifting the fundamental wavelength further into the red used to be Raman shifting, for example, with hydrogen gas. Only a very limited number of discrete lines were available by this technique. The major problem with these fundamental wavelengths further in the red is the reduced sensitivity of photocathode materials for their second-harmonic wavelength. By means of optical parametric oscillators, it is now possible to extend the range of the fundamental wavelength in a continuous way, both in the visible and in the infrared. Since the experiment itself is a nonlinear optical technique, and since the optical techniques to generate alternate fundamental wavelengths also rely on nonlinear optical techniques, a smooth temporal and spatial profile for the laser pulse is highly desirable. Injection seeding of the YAG slave laser by a diode pumped, frequency-stabilized seed master laser is strongly advised.

The low duty cycle of these nanosecond pulsed lasers requires the use of gated integrators (boxcar averagers). This allows for substantially increasing the signal-to-noise ratio of these experiments. Such lasers are also run at constant thermal load. Since the primary experiment is measuring the incoherently scattered light intensity at the second-harmonic wavelength as a function of fundamental light intensity, it is essential to be able to vary this intensity externally. This is realized by a half-wave plate between two polarizers. The first polarizer ensures complete polarization of the laser input, which is essential for obtaining 100% intensity modulation. The half-wave plate rotates the plane of polarization of the laser light beam, whereas the second polarizer defines the exact polarization (vertical). By rotating the half-wave plate, the polarization of the fundamental is rotated from perpendicular to parallel with the vertical polarization as defined by the second polarizer. In this way, the incoming power is modulated from zero to maximum. A beam sampler splits off a small fraction of this power toward a monitoring fast photodiode. This diode is used for both pulse temporal analysis and for measuring the relative fundamental power. The remainder of the laser beam is then focused in an optical cell. Due to the high power levels that are being used, it is essential to ensure the use of "high-power optics" in the excitation path.

Also, because of the high peak intensity that is impinging on the cell, a special design for nanosecond HRS has been developed. Since any non-centrosymmetric structure in the focus might give rise to second-order nonlinear response, high intensities at the glass sides should be avoided. Therefore, a long cell design is used that allows for the largest part of the focusing in the solution itself. Originally, very slow focusing ($f = 20$ cm) was used to minimize differences in focal position for different solvents, due to their different refractive indices. With the idea of performing half of the focusing in the solutions, cell lengths of 20 cm were required. Together with the substantial beam diameter of Q-switched lasers, this resulted in cell volumes that were excessive. Especially in view of the screening of precious new materials that are only available in milligram quantities, a much smaller and more practical cell design is now in use. Also, for the sake of combining this HRS technique with in situ electrochemistry (see Section 2.4.), it was essential to reduce the volume of the HRS cell, so as to lower the time required for a complete electrolysis.

Since the HRS intensity measurement itself involves the detection of very low levels of scattered photons, the detection path consists of an efficient photon condenser system and a photomultiplier tube. The condenser system is comprised of a retro-reflecting concave mirror: a large numerical-aperture aspheric lens to collect as many photons as possible, and a plano-convex lens to direct and loosely focus them on the photocathode. The photomultiplier has the appropriate interference filter fitted in its housing. The fundamental intensity variation (by means of the rotation of the half-wave plate) and the data collection are computer controlled.

For HRS intensity measurements, the experiment consists of recording the scattered light intensity at the second-harmonic wavelength as a function of incident fundamental laser light intensity. For a single molecular scatterer, the scattered intensity at the harmonic wavelength can be calculated by performing the orientational average over all possible orientations (*vide supra*) for β:

$$I_{2\omega} = \frac{16\,\pi^5}{c\,\lambda^4\,r^2}\left\langle \beta_{HRS}^2 \right\rangle I_\omega^2 \tag{2.19}$$

Keeping in mind that these individual scatterers do not bear any specific phase relations with respect to one another (incoherent scheme), we arrive at the expression for an ensemble of N molecular scatterers in a particular medium, or, in chemical terms, for N solute molecules in a particular solvent:

$$I_{2\omega} = \frac{16\,\pi^5}{c\,\lambda^4\,r^2}\,N\,f_\omega^4\,f_{2\omega}^2\left\langle \beta_{HRS}^2 \right\rangle I_\omega^2 \tag{2.20}$$

where N is the concentration or number density of the chromophores, and f_ω and $f_{2\omega}$ are the local field factors, as defined in Chapter 1.

For a solution of two components (solvent and solute) where both are noncentrosymmetric and both contribute to the HRS signal, the harmonic intensity $I_{2\omega}$ originates from the scattering by both the solute and the solvent molecules:

$$I_{2\omega} = G\left(N_s\left\langle \beta_{HRS}^2 \right\rangle_s + N_x\left\langle \beta_{HRS}^2 \right\rangle_x\right) I_\omega^2 \tag{2.21}$$

where G includes not only the wavelength λ and distance to the detector r, but also all experimental factors such as the photon collection efficiency and the coherence parameters for the fundamental laser. The subscripts s and x refer to solvent and chromophores, respectively. From a concentration series, the unknown $\langle \beta_{HRS}^2 \rangle_x$ for the solute can be determined when $\langle \beta_{HRS}^2 \rangle_s$ is known for a particular solvent. This method is referred to as the internal reference method. This method is only applicable for noncentrosymmetric solvents that result in an *intercept* for the quadratic coefficient (Q.C.):

$$Q.C. = \frac{I_{2\omega}}{I_\omega^2} = G N_s \langle \beta_{HRS}^2 \rangle_s + G \langle \beta_{HRS}^2 \rangle_x N_x = intercept + slope.X \quad (2.22)$$

which is comparable to the *slope*. Note how the instrumental factor G cancels out in both *intercept* and *slope*, allowing the retrieval of $\langle \beta_{HRS}^2 \rangle_x$ from a concentration series of the solute x. Although $\langle \beta_{HRS}^2 \rangle_s$ for a particular solvent might be relatively small, for small and mostly saturated molecules indeed, it is the number density of these molecules in the solvent, approximated by the constant concentration in the neat liquid for concentrations of the solute that are at most millimolar, which favors a large *intercept* $N_s \langle \beta_{HRS}^2 \rangle_s$.

In cases when (1) the solvent is noncentrosymmetric, but does not have a large enough $\langle \beta_{HRS}^2 \rangle_s$, or (2) the solvent is centrosymmetric, it is possible to apply the external reference method, which uses an external reference solute molecule with known $\langle \beta_{HRS}^2 \rangle_{ref}$. From a concentration series of the reference molecule and a concentration series of the unknown, and by comparison of both slopes, the unknown can be deduced:

$$Q.C._{ref} = \frac{I_{2\omega}}{I_\omega^2} = G \langle \beta_{HRS}^2 \rangle_{ref} N_{ref} = slope_{ref} X \quad (2.23)$$

$$Q.C._x = \frac{I_{2\omega}}{I_\omega^2} = G \langle \beta_{HRS}^2 \rangle_x N_x = slope_x X \quad (2.24)$$

$$\frac{slope_x}{slope_{ref}} = \frac{\langle \beta_{HRS}^2 \rangle_x}{\langle \beta_{HRS}^2 \rangle_{ref}} \quad (2.25)$$

With the primary experiment being a measurement of the dependence of the second-harmonic intensity as a function of fundamental intensity, it is essential to ensure that the collected signal is purely second-order in nature. Therefore, an interference filter for the second-harmonic wavelength is put in front of the photomultiplier. In principle, this should discriminate in the spectral domain between useful signal and useless stray light. However, it has been pointed out that apart from pure HRS, another source of incoherent photons at exactly the second-harmonic wavelength is possible. This is multiphoton fluorescence (either Stokes or anti-Stokes two-photon or three-photon fluorescence; Flipse et al. 1995).

Because this can be an additional contribution to the HRS signal, it can result in a systematic overestimation of the first hyperpolarizability. Multiphoton fluorescence (MPF) is an odd-order nonlinear optical property, two-photon fluorescence being related to the imaginary part of the third-order hyperpolarizability. However, the intensity dependence is also quadratic, so a pure quadratic curve does not guarantee a pure HRS signal. Moreover, since the multiphoton fluorescence contribution can be at exactly the second-harmonic wavelength, the use of interference filters does not guarantee either pure HRS response. Therefore, one has to revert to other techniques to (1) check for the presence of an MPF contribution to the HRS signal; and, if so, (2) correct for the overestimation in hyperpolarizability that would result from it.

In the spectral domain, a differentiation between the broadband MPF and the narrowband HRS is readily obtained. With MPF present, the HRS peak presents itself as a narrow peak, with its width determined by the spectrometer's spectral resolution (a few wavenumbers usually), superimposed on a broad continuous MPF background (Song et al. 1996). Essentially the same can be obtained with a discrete set of interference filters: If only a signal is obtained at exactly the second-harmonic wavelength and no signal is leaking through at other discrete wavelengths, the signal is pure HRS. If not, a technique of spectral background subtraction is required.

In the time domain, it may be intuitive to differentiate between the immediate (nonlinear) scattering and the time-delayed (multiphoton) fluorescence. By appropriate use of the time-gating of the boxcar averagers, differentiating between HRS and MPF might be obtained (Noordman and van Hulst 1996). In the time domain, if only a signal is obtained during the fundamental laser pulse, no time-delayed fluorescence is observed and the signal is pure HRS. If not, a technique of varying the time delay for the measurement time window in combination with an elaborate analysis scheme is in order. Also, since typical fluorescence lifetimes are in the nanosecond regime, shorter pulse duration for the fundamental is essential.

It may be less intuitive to see that in the frequency domain, there is a very attractive way of (1) showing the presence of an MPF contribution to the HRS signal, and (2) arriving at a precise and accurate hyperpolarizability value that is not overestimated by this MPF. The frequency domain is the Fourier transform of the time-domain technique. By measuring the modulus of the hyperpolarizability in the frequency domain, for increasing amplitude modulation frequencies, it is possible to measure the MPF affected hyperpolarizabilities for low amplitude modulation frequencies and to observe the lowering of this hyperpolarizability for high AM frequencies when the slow fluorescence response can no longer follow the fast excitation (Olbrechts et al. 1998). This demodulation ends for high frequencies at the MPF-free accurate hyperpolarizability value. It is possible to determine at the same time the phase angle between real and imaginary parts of the Fourier transforms. By simultaneous analysis of the experimental modulus and phase data toward the MPF-free true hyperpolarizability, the MPF contribution, and the fluorescence lifetime, it is possible to arrive at a very precise estimate of all three values (Clays et al. 2001a, 2001b).

Experimentally, the amplitude modulation is obtained as the intrinsic higher harmonics of the femtosecond-pulsed laser with its repetition rate around the 100 MHz. Because of the extremely short duration of these pulses, the harmonic content extends well beyond the gigahertz regime, far beyond the bandwidth of the photomultiplier detectors. The generic laser for the femtosecond pulses is the titanium-sapphire laser, generating approximately 100 femtosecond pulses and tunable around 800 nm. This is also a convenient fundamental wavelength, which in turn can also be shifted by parametric oscillation. In this way, the two near-infrared wavelengths that are relevant for glass fiber optical telecommunications can be realized (1300 and 1550 nm).

Because of the much higher repetition rate for these lasers (when not amplified), a much higher duty cycle is obtained with integrating photomultipliers. Therefore, phase-sensitive detectors (lock-in detectors) are used rather than the gated integrators that are ideal for low duty cycle nanosecond-pulsed lasers. Also different from the nanosecond HRS with high peak power pulses is the cell design. Associated with the higher repetition rate is the lower peak power in these femtosecond pulses. As a result, commercially available 1-by-1 cm² spectroscopic cells can be used. Finally, since the laser beam is much more confined, it is possible to use 2 mm wide cells, substantially reducing the total cell volume. This has advantages in terms of the amount of product needed, as well as in terms of time needed for electrolysis (see Section 2.4).

The remainder of the optical setup and electronics is very similar. Again, a half-wave plate between polarizers is used to modulate the fundamental beam intensity. Now, special care has to be taken to preserve the femtosecond duration of the pulse. This requires the use of so-called "femtosecond optics" with zero-group velocity dispersion around the design wavelength in the excitation path.

For the experimental determination of the HRS depolarization ratio, an additional analyzing polarizer is inserted in the detection path. In principle, this depolarization ratio is to be determined in the limit of zero numerical aperture (Morrison et al. 1996). Therefore, if the signal strength allows, the condenser system is removed. Otherwise, a variable aperture is inserted and the minimal aperture that still allows the signal to be detected is used. Other factors determining the observed depolarization ratio are the relative contribution from the solvent and from the fluorescence. Up to now, we have assumed in the analysis of the depolarization ratio that the signal is completely and only determined by the solute for which the symmetry is being determined from its depolarization ratio. However, for a noncentrosymmetric solvent, there is nonzero solvent contribution to the signal. If the solute is contributing very little to the signal, that is, because its hyperpolarizability is very small or its concentration is low, the depolarization ratio will largely reflect the symmetry of the solvent, rather than that of the solute under investigation. Otherwise, one has to determine the quadratic coefficient (Q.C.; Equation 2.22) for the vertical (Z) and horizontal (X) components of the HRS light versus the concentration of the solute x:

$$Q.C._Z = G N_s \langle \beta^2_{ZZZ} \rangle_s + G \langle \beta^2_{ZZZ} \rangle_x N_x \qquad (2.26)$$

$$Q.C._X = G N_s \langle \beta^2_{XZZ} \rangle_s + G \langle \beta^2_{XZZ} \rangle_x N_x \qquad (2.27)$$

The ratio of the intercepts for both polarizations will yield the depolarization ratio of the solvent and the ratio of the intercepts will give the depolarization ratio of the solute.

A second artifact in depolarization measurements is, again, multiphoton fluorescence. Although it has been shown that, at least for dipolar and octupolar symmetry, the depolarization ratio for HRS and MPF are similar, this analysis overlooks the very different dynamics for these two processes. Although HRS is instantaneous and, hence, does not allow time for rotational relaxation, MPF is largely time-delayed, allowing for at least partial rotational relaxation between excitation and emission. In linear fluorescence studies, this phenomenon allows for the determination of the rotational relaxation time. What is essential for the influence on the HRS depolarization ratio is that for long enough fluorescence lifetimes, the time between excitation and emission allows for complete depolarization of a signal that is largely comprised of fluorescence. Depolarization ratios of 1 have been determined for fluorescent dipolar species that should exhibit a ratio of at least 1/3.

It is interesting that, at one instance, a depolarization ratio has been determined as a function of AM frequency in the frequency domain and with femtosecond pulses (Clays et al. 2003). Congruent with the rotational relaxation concept, an increase in depolarization ratio has been observed for higher AM frequencies, in agreement with the smaller fluorescence contribution for increasing frequencies. Unfortunately, a high-frequency limit could not be obtained. This is a very difficult experiment, since the signal amplitude inherently is reduced due to the demodulation of the fluorescent contribution at high frequencies.

2.3   Determination of molecular symmetries

Before coming to the application of HRS specifically for the determination of molecular symmetries, and to additionally appreciate this possibility, it is instructive to realize the added value of a second-order nonlinear technique, especially in tandem with a first- or third-order technique (or, in more general terms, the combination of an odd- and even-order technique). As already demonstrated, by the application of Neumann's principle, for an even-order technique to be applicable, noncentrosymmetry is a prerequisite. This condition strictly holds in the electric-dipole approximation. There is no such condition for odd-order techniques. Therefore, the combination of an odd-order (no condition) with an even-order technique (noncentrosymmetry condition) reveals the combined information of a general nature (like presence) in combination with symmetry information (noncentrosymmetry when a signal is observed, centrosymmetry when no odd-order response is found).

A particularly nice example of this combination was demonstrated with amphiphilic nonlinear optical chromophores (NLOphores) dispersed in membranes for nonlinear optical microscopy (Moreaux et al. 2000). In addition, the odd-order optical response was not first-order linear but third-order nonlinear to enhance the spatial resolution in the depth dimension. The microscopic images obtained from two-photon absorption-induced fluorescence (the third-order nonlinear optical technique) show the general information (presence of chromophores, by the virtue of their function as fluorophores) as the location of the membranes. The images obtained from second-harmonic generation (clearly, the second-order nonlinear optical technique) show only the locations characterized by noncentrosymmetry. To be somewhat more precise, SHG is generated only by the dispersed chromophores, by the virtue of their function of second-order NLOphores with large second-order polarizability, when the relative orientation of these moieties does not result in centrosymmetry. A free-standing membrane would show up in both images, two touching membranes with their flattened interface would show up completely in the fluorescence image, but the flattened interface with centrosymmetric arrangement of the NLOphores in the adjoining membranes would not show up in the SHG image. Current investigations are under way to improve the nonlinearity (both second- and third-order) of the chromophores to enhance the sensitivity of this nonlinear microscopic technique. More strongly polarizable, noncentrosymmetric and amphiphilic molecules are being designed and evaluated for their second- and third-order polarizability to this purpose. Well-known multiphoton fluorescent emitters, such as green or red fluorescent proteins (GFP or RFP) are also being investigated for labeling and subsequent nonlinear imaging.

Although the aforementioned nonlinear microscopy does not pertain to incoherent HRS, but rather to coherent SHG, it gives an idea about the potential of HRS for structure determination: If an (true, second-order nonlinear) HRS signal is observed (not convoluted by third-order MPF) from a molecular scatterer, its structure should be noncentrosymmetric of some sort. For single-oscillator dipolar molecules, it is trivial to derive the approximate symmetry. The fruit fly of nonlinear optics, paranitroaniline (PNA), is an instructive example. However, it is again essential to combine odd-order linear optics with the even-order HRS technique, to observe whether there is also "effective optical noncentrosymmetry." We coined this term because there are also clearly noncentrosymmetric molecular structures that do not generate any second-order nonlinear response because the second-order polarizability is still negligible. This is the case when the polarizable core of the molecule is substituted by two clearly different moieties (to arrive at the "crystallographic noncentrosymmetry") but with similar electronic properties (the same tendency to donate or withdraw electrons, or, in chemical terms, the same acid-base or redox properties). Reverting to paranitroaniline, protonation of the amine group to arrive at the polarizable benzene core substitute with two different moieties ($-NH_3^+$ and $-NO_2$) clearly results in still a noncentrosymmetric structure, but the original combination of the electron donor with the electron acceptor has now turned into a second-order nonlinear optically dead, yet noncentrosymmetric structure. The protonated amine group is an electron acceptor, same as the nitro group. In the linear optical regime, there is no charge-transfer absorption band for the two substituents having the same tendency to donate or withdraw electrons, since there is no charge transfer between donor and acceptor. It is interesting to note how there is, to our knowledge, no experimental demonstration of this *gedanken* experiment for paranitroaniline, but there is for the similar molecule with acceptor C60 fullerene rather than the nitro group (Asselberghs et al. 2002).

Although this may seem straightforward, it is essential to combine linear with second-order linear optical experiments to arrive at the correct conclusions for molecules more elaborated than the prototypical single donor-π-acceptor systems. Correlated chromophores in a molecular superstructure are a case in point. When from linear absorption it is observed that (1) the charge transfer is still present, but more important, (2) that the electronic properties of the individual chromophores in the superstructure remain unchanged, it is possible to arrive at rather detailed conclusions about the symmetry of these superstructures. Because the individual chromophores are (small, microscopic) molecules, and the bulk (macroscopic) material is a solution or ensemble, not of the small molecules, but of the intermediate superstructures, the latter have also been referred to as mesoscopic (not be confused with mesogenic).

To recapitulate from Section 2.2, it is only the combination of an HRS intensity measurement

$$\left\langle \beta^2_{HRS} \right\rangle = \frac{6}{35}\sum_i \beta^2_{iii} + \frac{16}{105}\sum_{i \neq j}\beta_{iii}\beta_{ijj} + \frac{38}{105}\sum_{i \neq j}\beta^2_{iij} + \frac{16}{105}\sum_{ijk,cyclic}\beta_{iij}\beta_{jkk} + \frac{20}{35}\beta^2_{ijk}$$

(2.28)

with an HRS depolarization ratio measurement

$$\rho = \frac{\left\langle \beta^2_{XZZ} \right\rangle}{\left\langle \beta^2_{ZZZ} \right\rangle}$$

(2.29)

that allows for the analysis toward the value for a molecular hyperpolarizability tensor component. In principle, from the two independent experimental observables—the total HRS-intensity-derived $\left\langle \beta^2_{HRS} \right\rangle^{1/2}$ and the ratio-derived depolarization ratio—two independent hyperpolarizability tensor components can be deduced.

In actual practice, it is experimentally difficult to measure an accurate depolarization ratio, as already mentioned. First, the low efficiency of HRS is even further reduced by inserting an analyzing polarizer. Second, to compensate for the low efficiency of photon collection, a condensing system with large numerical aperture is used, but this results in a large solid angle for detection, and, hence, a less exactly defined or "scrambled" polarization for the detected photons. This results in a lowering of the depolarization ratio with larger numerical aperture. Obviously, reducing the numerical aperture is the way to alleviate this problem, but this additionally lowers the signal-to-noise ratio. Finally, as already discussed, any multiphoton fluorescence contribution at low amplitude modulation frequency will also lower the observed depolarization ratio. Experimental HRS depolarization ratios of 1 at low AM frequency have been reported, but they have been clearly shown to be AM frequency dependent, increasing to the theoretical value for the symmetry in case.

It is essential to appreciate how these different contributions to the depolarization ratio might affect the result. Most often, the experimental values are lower limits, since a nonzero numerical aperture or a not completely demodulated fluorescence contribution and the concomitant nonideal high-frequency extrapolation for the depolarization ratio, tend to lower the value. Finally, for the prototypical one-dimensional (1-D) elongated donor-acceptor molecule with the single major diagonal tensor component β_{zzz}, the depolarization ratio (ρ) is identical to

$$\rho = \frac{\frac{1}{35}\beta_{zzz}^2}{\frac{1}{7}\beta_{zzz}^2} = \frac{1}{5} \tag{2.30}$$

but this is only in the mathematical 1-D limit of zero (hyper)polarizability in the other dimensions.

With all the aforementioned experimental intricacies and theoretical considerations, for molecules like para-nitroaniline (pNA) and disperse red 1 (DR1; two typical dipolar reference molecules), depolarization ratios of typically 0.33 to 0.28 have been experimentally obtained and are considered as signatures of a strongly dipolar nature of the nonlinear optical response.

At the other extreme, depolarization values significantly higher than 2/3 should not be considered, since 2/3 is the theoretical higher limit for this ratio (pure octupolar response). A large numerical aperture will not strongly affect (lower) this value but fluorescence might.

However, values for the depolarization ratio between 2/3 and 2/5 are hard to correctly interpret, even if the value is not due to a finite aperture or to a fluorescence contribution. There are a number of molecular symmetries that give rise to a higher depolarization ratio.

Most clearly, purely octupolar symmetries result in a depolarization ratio of 2/3 exactly. This pertains to the molecular noncentrosymmetric D_{3h} and T_d symmetries and to the less well-known D_2 and D_{2d} symmetries. The former two symmetries are particularly well known in materials science. The first is often the consequence of threefold symmetric substitution of planar conjugated sp^2 systems, and a 1,3,5-trisubstituted benzene ring is the prototype (Verbiest et al. 1994). It can result in discotic mesogens with interesting liquid crystalline properties, which can self-assemble in noncentrosymmetric mesophases (Hennrich et al. 2006). From symmetry arguments, if follows that the only nonzero and dependent tensor components are equal to $\beta_{zzz} = -\beta_{zxx} = -\beta_{xxz} = -\beta_{xzx}$ (Verbiest et al. 1993). Tetrahedral symmetry often results from fourfold symmetric substitution on a saturated sp^3 carbon atom. CCl_4 is the historic example and the only nonzero tensor element was reported in 1965 reported to be β_{xyz} (Terhune et al. 1965). Because of the clear molecular topology that is associated with these symmetries, depolarization ratios are often only used to confirm what has already been derived from the molecular structural formula.

Quite a different situation results in two cases. One is when the molecular structure is clearly not of octupolar symmetry and the depolarization ratio is high (but still lower than 2/3, to be technically correct); the other is when a molecular structure might seem rather centrosymmetric, but when significant second-order scattering is observed, indicating actual noncentrosymmetric symmetry.

The first case (high depolarization ratio for dipolar structures) can result from a nonzero off-diagonal hyperpolarizability tensor component, contributing to the HRS response. For example, if we take a molecule with C_{2v} symmetry with two independent components—β_{zzz} (along the molecular dipolar z-axis) and the off-diagonal β_{zxx} tensor component—the total HRS intensity becomes a function of both tensor components:

$$\langle \beta_{HRS}^2 \rangle = \frac{6}{35}\beta_{zzz}^2 + \frac{16}{105}\beta_{zzz}\beta_{zxx} + \frac{38}{105}\beta_{zxx}^2 \tag{2.31}$$

and the depolarization ratio (ρ) can be rewritten as a function of the molecular ratio q between off-diagonal and diagonal tensor components, $q = \beta_{zxx}/\beta_{zzz}$:

$$\rho = \frac{3 - 2q + 11q^2}{15 + 18q + 27q^2} \tag{2.32}$$

Careful inspection of this relation shows that when the off-diagonal tensor element β_{zxx} equals zero, ρ reverts to the limiting value of 1/5 and

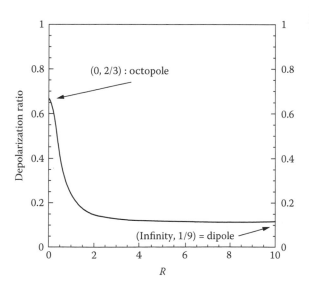

Figure 2.5 Depolarization ratio versus R.

equals $2/3$ for $q = \beta_{zxx}/\beta_{zzz} = -1$. In general, however, a wide range of values for the depolarization ratio is possible.

Alternatively, one could rewrite Equations 2.2, 2.3, and 2.28 as a function of the dipolar and octupolar hyperpolarizabilities (see Chapter 1, Section 1.4) as:

$$\left\langle \beta_{ZZZ}^2 \right\rangle = \frac{9}{45}\left|3\beta_3^1\right|^2 + \frac{6}{105}\left|\beta_3^3\right|^2 \tag{2.33}$$

$$\left\langle \beta_{XZZ}^2 \right\rangle = \frac{1}{45}\left|3\beta_3^1\right|^2 + \frac{4}{105}\left|\beta_3^3\right|^2 \tag{2.34}$$

$$\left\langle \beta_{HRS}^2 \right\rangle = \frac{10}{45}\left|3\beta_3^1\right|^2 + \frac{10}{105}\left|\beta_3^3\right|^2 \tag{2.35}$$

where $||$ represents the norm of the tensor, that is, the square root of the square of all tensor components.

The depolarization ratio can than be defined as

$$\rho = \frac{7R^2 + 12}{63R^2 + 18} \tag{2.36}$$

where $R = |3\beta_3^1|/|\beta_3^3|$. It varies between $2/3$ (pure octupole: $\beta_3^1 = 0$) and a limiting value of $1/9$ when $|R| \to \infty$ as shown in Figure 2.5. This last case could be considered as a pure dipole, but in reality every dipolar molecule will also exhibit an octupolar contribution.

Another case in point is seemingly centrosymmetric molecular structures, when only considering two-dimensional representations (such as when written on paper, printed in books, or schemes on the blackboard, often luring students to assume centrosymmetry). But when appreciating the third dimension and realizing possible torsional motion, possible partly breaking planar conjugation, very exciting alternative octupolar symmetries might arise, namely, D_2 and D_{2d} symmetries. An example of such a structure is shown in Figure 2.6 (Duncan et al. 2008).

It is only when considering torsional motion out of the plane that an actual noncentrosymmetric structure arises, amenable of generating second-order nonlinear optical response. For such a highly symmetric (but not a centrosymmetric structure), it is first essential to demonstrate that there be no odd-order fluorescence contribution to the HRS signal. This can be positively ascertained by demonstrating the absence of any high-frequency demodulation to the HRS signal. Then, the actual

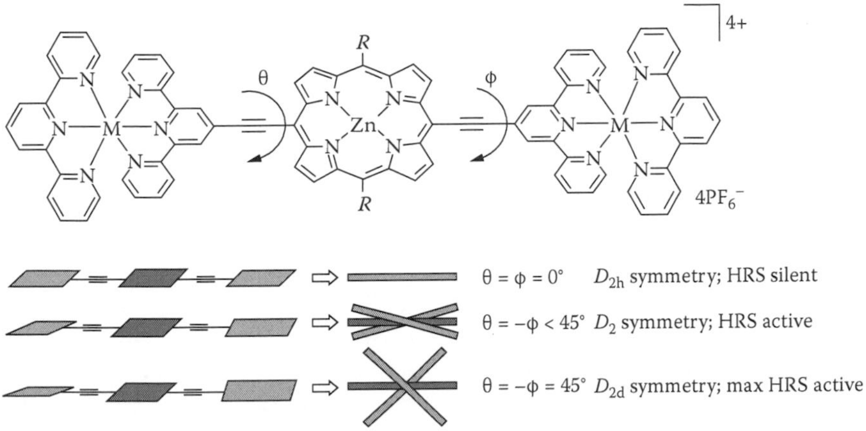

Figure 2.6 Schematic representation of coupled oscillators with effective D_2 non-centrosymmetry derived from torsional motion (Duncan et al. 2008).

symmetry of noncentrosymmetric structures can be assessed by the depolarization ratio. The experimental values between 2/3 and 1/2 show that the symmetry should be either dipolar with a large off-diagonal tensor component β_{zxx} (*vide supra*) or octupolar with a small dipolar contribution. Clearly, from the chemical topology, a dipolar symmetry with $\beta_{zxx} = -\beta_{zzz}$ is precluded, as are octupolar structures with D_{3h} and T_d symmetries. Therefore, we have concluded the actual symmetry for the compound shown in Figure 2.6 to be D_2, or in the limiting case of $\theta = \phi$ equal to 45°, D_{2d}.

These examples show that it is possible to derive important information from the combined intensity and depolarization HRS measurements. The intensity measurement $\langle \beta^2_{HRS} \rangle^{1/2}$ determines if a molecular structure is noncentrosymmetric, that is, showing second-order nonlinear response. It is crucial to ensure complete absence of odd-order (one- or two-photon fluorescence, linear, and third-order nonlinear) contributions to the response, since these are allowed for centrosymmetric structures. After establishing noncentrosymmetry without any doubt from HRS intensity measurements, the actual symmetry can be derived with the aid of the HRS depolarization ratio, in combination with chemical topology.

Finally, it is important to appreciate that this symmetry is determined in solution and at room temperature. On the one hand this means that the structure is dynamic and that the corresponding symmetry is an average result. It is the structure that is relevant for chemical reactivity. On the other hand, this also means that a molecular structure can be determined without the need for x-ray diffraction. Not all molecules are amenable to single-crystal formation and the rigid structure in the solid state might not be relevant for dynamic solution phase properties and chemical reactivity.

2.4   Switching the first hyperpolarizability at the molecular level

One of the specific advantages of the incoherent scattering nature of the HRS technique is its applicability to ionic species. This allows for the study of the switching of the second-order nonlinear optical properties at the molecular level. This particular kind of switching is very different from the switching that is realized in, for example, an electro-optic switch. In such a switch, or modulator, an applied electric field is modulating the effective refractive index of the material in the modulator, but the molecular material itself remains unchanged. It is the particular nonlinear effect, in this case the Pockels effect, that causes an electric field to change the refractive index of a material. The material itself, or the optical or other properties of the molecular material, does not change.

Switching at the molecular level, on the other hand, does involve changing of the molecular material itself. This kind of switching involves chemical reactions between two distinctive and relatively stable states. For such reactions to have an effect on the optical properties of the molecules, it is important to recall what molecular moieties do to impart optical properties (in the visible spectrum, color) to molecules.

In general, for linear and all order nonlinear optical properties, highly polarizable electrons are needed for the necessary interaction with the varying electromagnetic field of the light. Such polarizable electrons are found in organic conjugated systems. The corresponding systems typically consist of carbon atoms in sp^2 hybridization, resulting in planar molecular structure where a benzyl-like moiety is involved with bond angles of 120°. Following the particle-in-a-box concept (Atkins 1978), the more extended the conjugated molecule, the more the absorption spectrum shifts toward lower energy (bathochromism effect), that is, the longer the maximum wavelength absorption. This means that a change in nature between conjugated and saturated bonds (carbon atoms in sp^3 hybridization, with bond angles of 109.5°, hence out of a single plane), or a change in length of the conjugated systems, will result in a color change (modification of the energy gap). An equivalent effect is observed in NLO: the bond length alternation (BLA) criterion in conjugated systems improves the quadratic NLO efficiency, for instance (Marder et al. 1994; Champagne and Bishop 2003).

However, for any even-order nonlinear optical effect to be switched there is a need of a change in the degree of noncentrosymmetry in the NLO response. This translates into changing the electron donating strength of a donor or the electron withdrawing strength of an acceptor. There are a number of chemical means to do this. These are related to the chemical reactivity that is to be associated with an electron rich moiety in a molecule. Such a moiety, being rich on negative electrons, is prone to donate electrons over the conjugated, strongly polarizable sp^2 carbon system to a more electron-deficient moiety in the same molecule. This gives rise to the so-called charge-transfer absorption band. This very same electron-rich moiety is also prone to chemical attack by another electron-deficient molecular species, with the aim of creating a new chemical bond. Examples of such kinds of species are positively charged metal cations and the proton (acido-basic reactions). When metal cations react with electron-rich molecules, the latter are called ligands. These ligands then form a complex ion with a central metal cation in a type of chemical reaction that is generically referred to as complexation. The same kind of reaction can occur with the proton in proton-transfer or acid-base reactions. As a result of complexation or protonation of the electron-rich moiety, this is now no longer so electron-rich, and its nature has changed from electron donating to rather electron withdrawing.

Considering the realization of the noncentrosymmetry requirement for second-order nonlinear optics, this complexation or protonation means a lowering of the degree of noncentrosymmetry in a donor–acceptor charge-transfer couple. Please note that a molecule is structurally either centrosymmetric or noncentrosymmetric, and that no intermediate situation is plausible. However, for the optical effect of the structure, a molecule with two structurally different moieties on both ends of its conjugated bridge, but with identical electron accepting or withdrawing capacity, will effectively be centrosymmetric and will not exhibit an charge-transfer optical absorption band. Therefore, we like to introduce the concept of "effective symmetry" for its optical relevance.

This effective symmetry may resemble the approximate symmetry that is often used when only considering the conjugated sp^2 hybridized molecular plane, irrespective of the saturated sp^3 substituents that are covalently linked to this plane. Although the exact molecular symmetry is lowered by the presence of these substituents, for the optical properties (determined by the charge transfer over the molecular plane) only the higher, approximate symmetry is relevant. However, the concept of effective symmetry goes beyond this approximate symmetry. Also in the approximation of the simple molecular plane, a structurally noncentrosymmetric moiety may turn out to be optically effectively centrosymmetric. A simple example may help here, and we can again use our fruit fly of the field: paranitroaniline is noncentrosymmetric, with the amine donor and the nitro acceptor group at opposite ends of the strongly polarizable and conjugated benzene ring. Therefore, paranitroaniline exhibits a charge-transfer absorption band, and, by the same token, a sizeable first hyperpolarizability. However, upon protonation of the amine donor group (which is very easy with a protic solvent), the resulting positively charged ammonium group is now also an accepting group, and the structure is still strictly noncentrosymmetric, with now an ammonium and a nitro group at opposite ends, but the central benzene core is now feeling two electron acceptor groups. Hence, there is no charge-transfer absorption band and negligible second-order nonlinearity. Or, the molecule is now "effectively centrosymmetric."

This has experimentally been demonstrated, be it on a more complex molecular system, with a C60 fullerene as the electron deficient moiety and the same amine group as electron donor, both on benzene (Giraud et al. 2007). Clearly, whether the amine base is protonated is not going to change the exact noncentrosymmetry of the large molecule. However, the neutral amine induces a charge-transfer absorption band indeed in conjunction with the fullerene, whereas the protonated ammonium does not. In our concept of effective centrosymmetry the ammonium group and the fullerene ball are optically equivalent and results in an effective centrosymmetric structure without the charge-transfer band and without second-order optical nonlinearity.

Clearly, HRS is the only technique capable of experimentally determining the lowering of the first hyperpolarizability of the protonated species. Such an ionic species is not amenable of electric-field induced orientation, since it would rather migrate than orient.

It may have occurred to the reader that the optically relevant properties of electron donating and accepting powers are inherently related to the chemical properties of bases and acids, respectively. Indeed, proton transfer is a convenient way to turn bases into their "conjugated" acids. Please note that the term *conjugate* does not have the same meaning as used previously in the context of neighboring sp^2 carbon atoms. Here, this term refers to the acid-base couples that are inherently related to each other (conjugated) through the proton transfer. As a consequence of the acid-base properties of electron-accepting and -donating groups, respectively, protonation and deprotonation can be used to switch the linear and nonlinear optical properties of chromophores where the noncentrosymmetry has been induced by bases and acids. Apart from the demonstration of principle already mentioned for a simple approximately linear donor–acceptor system, protonation has also been used in more elaborate organometallic complex ions (Coe et al. 2006) and in octupolar species.

A quite interesting elaboration of the protonation effect was realized in an octupolar system (Asselberghs et al. 2006). When the octupolar molecule consists of arms that are amenable to protonation or deprotonation, the octupolar first hyperpolarizability can be switched by proton transfer also. In principle, a transition from octupolar for all branches in their neutral form, over dipolar for only a subset of branches protonated, to octupolar again for all branches protonated should be observable. However, experimentally, for D_{3h} trigonal octupoles, such a transition was not observed. But a strong enhancement of the second-order nonlinear response of D_{3h} octupoles with an electron-rich central benzene core was observed upon protonation of the three pyridine base branches to pyridinium acceptors. Concomitant with the enhanced nonlinear response, a bathochromic shift of the electronic absorption was observed, congruent with an increased charge-transfer feature in the absorption spectrum.

Although the protonation–deprotonation scheme offers a convenient and reversible switching scheme based on altering the pH of a solution, it is not so amenable to fast switching in a thin film format, as often envisaged for photonic applications. Also direct electronic control of the switching based on changing the pH is problematic.

However, just as the optically relevant properties of electron donating and accepting powers are inherently related to the chemical properties of bases and acids, respectively, similarly they are also related to the chemical properties of reductant and oxidant, respectively. A reductant is a reagent capable of reducing another reagent, by transferring electrons from the reductant to that reagent, so as to lower ("reduce") the oxidations stage of an atom in that reagent. Conversely, an oxidant is capable of increasing the oxidation stage of an atom in another reagent by withdrawing electrons from that reagent and taking those electrons up. Therefore, a reductant as a moiety on a conjugated part acts as an electron donor. Moreover, in an electron-transfer chemical reaction, it may be oxidized to become an oxidant itself, thereby changing from electron-donating to -accepting moiety.

Typical examples of electron-transfer amenable moieties are transition metal complexes. The term *transition* already refers to the different oxidation states that such metals can adopt. Well known are Fe(II), as in ferrocene, Ru(II), and Os(II) that can be easily and reversibly oxidized to Fe(III) (in ferrocenium), Ru(III), and Os(III), respectively. When used in conjunction with an electron-acceptor group, such as the nitro group, for example, on nitrothiophene, the donor–acceptor Fe(II)-nitro-based resulting non-centrosymmetric molecule can be oxidized to the "effectively" centrosymmetric Fe(III)-nitro acceptor–acceptor molecule with no charge-transfer absorption and zero first hyperpolarizability (Malaun, Kowallick, et al. 2001; Malaun, Reeves, et al. 2001; Asselberghs et al. 2003).

Ruthenium has proven to be extremely versatile as redox-amenable central metal cation in complexes for second-order nonlinear optics. When one of the six neutral spectator ligands is replaced by one with a long conjugated path that ends with an electron deficient moiety, extremely efficient structures can be obtained with a high first hyperpolarizability that can be reversibly switched by applying a potential. Advanced molecular engineering can be used to design a chromophore with a redox-switchable electron-donor group, with a conjugated part of variable length to tune the charge-transfer band for optimal resonance effects, with an electron-acceptor group to optimize the first hyperpolarizability, and with hydrophobic alkyl chains to impart amphiphilic properties to the chromophore (together with the hydrophilic redox-amenable polar group). Such a specifically designed molecule had been shown to exhibit redox-based reversible linear and second-order nonlinear optical properties as measured by HRS in solution, in combination with wet redox chemistry and dry electrochemistry (Coe et al. 1998). The same molecule was recently transferred by Langmuir–Blodgett surface transfer techniques to a solid thin-film format. In this much more practical solid-state format, it has been shown to exhibit the same second-order nonlinear optical properties that can be reversibly switched by electrochemistry (Boubekeur-Lecaque et al. 2008).

In conclusion, since the reviews by Coe (1999) and Delaire and Nakatani (2000), HRS switching behaviors have been investigated by experimental and theoretical means for many organic and organometallic molecular switches. Among these HRS switching systems that have been synthesized and/or subjected to experimental and theoretical characterizations are:

- Ruthenium(II)- and Fe(II)-based chromophores with reversible redox-switching HRS responses (Coe et al. 1999; Asselberghs et al. 2003; Coe 2006; Boubekeur-Lecaque et al. 2008)
- Substituted helicenes with large β enhancement upon oxidation (Botek et al. 2005, 2006)
- Push-pull bisboronate chromophore with a potential electric field induced NLO switching behavior (Lamère et al. 2006)
- Pyridine-based octupolar chromophores exhibiting β variations upon protonation (Asselberghs et al. 2006; Oliva et al. 2007)
- Photoswitchable dithiazolylethene-based derivatives incorporating a push-pull structural motif (Giraud et al. 2007)
- Photoswitchable zinc(II) complexes (Aubert et al. 2008)

It is also worth mentioning dipolar-type oxazolidine-based systems (Sanguinet et al. 2005, 2006; Mançois et al. 2006, 2007) that are photo- and/or acidochromic systems, commutating between an open and a closed form, the molecular second-order NLO response being about one order of magnitude larger for the open form due to a better electron delocalization (Figure 2.7). In conclusion, all these commutations can be triggered by a large set of external stimuli: pH, pressure, light, redox potential, electric field, magnetic field, T, ions, and so forth.

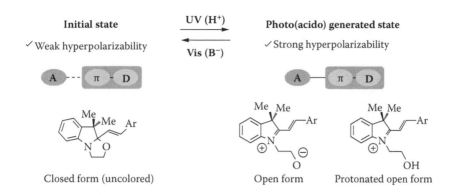

Figure 2.7 Photochromic/acidochromic equilibrium for the oxazolidine-based systems. A strong HRS contrast is observed between the uncolored form (closed) and colored open (or protonated open) form.

2.5 Probing aggregation and supramolecular structure in solution

Since HRS scattering is inherently sensitive to the symmetry of the scattering structure, it is also a useful tool to probe this symmetry in solution. Of course, there are other and much more advanced tools to obtain a more definite solution to the structure problem, but these techniques do not offer a solution-phase structure. HRS is quite unique in that it is capable of offering an idea about this very relevant information. High-resolution x-ray structures offer the ultimate in structural information, yet only on solid samples; much of the relevant properties (dynamics, reactivity) occur only in the solution phase.

To demonstrate the principle of using HRS to determine solution-phase structure, we start by analyzing the HRS response for correlated chromophores. These are defined as individual noncentrosymmetric chromophores with specific and time-invariant phase relations with respect to one another. These can be the consequence of covalent chemical bonds, as in carbazoles (Boutton 1998), binap (Hendrickx et al. 1997), or calixarene (Hennrich et al. 2005) compounds; or of protein assembly, as in bacteriorhodopsin (Clays et al. 1993). Also aggregation in solution can result in specific phase relation between chromophores. Dimerization is a simple case in point. For dimers of dipolar structures, the dipolar interaction energy results most often in centrosymmetric aggregates. In the other cases mentioned, there are multiple individual dipolar chromophores that are fixed with respect to one another in a supramolecular structure with noncentrosymmetry.

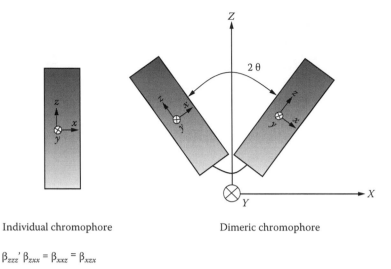

Individual chromophore   Dimeric chromophore

β_{zzz}, $\beta_{zxx} = \beta_{xxz} = \beta_{xzx}$

Figure 2.8 Individual chromophore arranged into a dimeric supramolecular structure.

In Figure 2.8 we illustrate this for a simple dipolar molecule with C_{2v} symmetry that is arranged into a dimeric structure. The hyperpolarizability of each individual chromophore is defined in the xyz coordinate system (with z along the charge-transfer axis), whereas the hyperpolarizability of the supramolecular structure is defined in the XYZ coordinate system. Now it is quite straightforward to write the hyperpolarizability of the dimer as a function of the hyperpolarizabilities of the individual chromophore. If we assume that the individual chromophore has only two independent components—β_{zzz} and $\beta_{zxx} = \beta_{xxz} = \beta_{xzx}$ (we assume the out-of-plane components β_{zyy} to be zero)—then we can calculate the non-vanishing components of the dimer in the ZXY coordinate system according to

$$\beta_{IJK}(\mathrm{dim}\,er) = 2\sum_{ijk} \beta_{ijk}(monomer)\cos(Ii)\cos(Jj)\cos(Kk) \qquad (2.37)$$

where $\cos(Ii)$, $\cos(Ji)$, and $\cos(Kk)$ are the direction cosines. This immediately yields the following hyperpolarizability components for the dimer:

$$\beta_{ZZZ} = 2\beta_{zzz}\cos^3\theta + 6\beta_{zxx}\cos\theta\sin^2\theta \qquad (2.38)$$

$$\beta_{ZXX} = \beta_{XZX} = \beta_{XXZ} = 2(\beta_{zzz} + 2\beta_{zxx})\cos\theta\sin^2\theta + 2\beta_{zxx}\cos^3\theta \qquad (2.39)$$

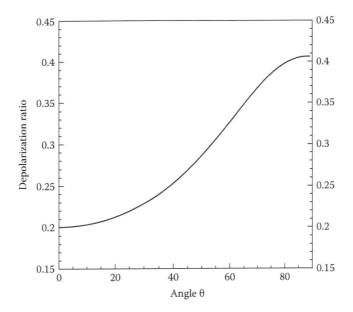

Figure 2.9 Depolarization ratio versus dihedral angle.

If we now know the values of β_{zzz} and β_{zxx} of the individual chromophore (for example, by a depolarization measurement), then we can determine the dihedral angle between the chromophore in the dimer by a depolarization measurement on the dimer. If we assume for simplicity that the zxx components are negligible compared to the zzz component, we obtain the depolarization ratio versus the dihedral angle as shown in Figure 2.9.

In the case of the neutral carbazoles, the known value for this angle could be retrieved from the combined HRS intensities and depolarization ratios, and, additionally, compared to the EFISH results. For the neutral binap compounds, a consisted picture in terms of dihedral angle between the two naphthalene planes could be obtained (Hendrickx et al. 1997).

The formation of dimers, mostly centrosymmetric aggregates, deserves our attention. This is a concentration-dependent phenomenon and can often already be studied by linear ultraviolet-visible absorption techniques. Since the HRS experiment also involves concentration-dependence, it can be instructive to do HRS measurements in a low-concentration regime and in a high-concentration regime. Aggregation can result in concentration-dependent values for the hyperpolarizability values. Most often, lower values are retrieved at higher concentrations, pointing to centrosymmetric dimer formation. Highly polar elongated structures are especially prone to this detrimental effect, which is enhanced in solvents of low polarity. Dimer formation can be a first indication of lower solubility, resulting in aggregation and ultimately precipitation.

The bacteriorhodopsin protein has been studied quite extensively by nonlinear optical techniques. Its two-photon absorption properties make it a good candidate for all-optical data manipulation. While it had been pointed out that the first hyperpolarizability of a protein cannot be measured by EFISH, its value had been deduced from its two-photon absorption spectrum and an estimate for the line width of this absorption (Birge and Zhang 1990). The reasons why bacteriorhodopsin is not amenable for EFISH experiments are its residual charges, even at the iso-electric point of the protein where there would not be migration under the influence of the electric field. The location of these charges would determine the orientation of the protein in the electric field, rather than the dipole moment of the chromophore embedded in the apo-protein. However, HRS is indeed ideally suited for the study of the second-order nonlinear optical properties of proteins. In addition, it turned out that HRS is particularly well suited, since this protein is available as purple membrane patches of the bacteriorhodopsin protein. The dynamics of the solubilization of these patches of multiple correlated protein molecules could be followed, both with second-order nonlinear optical techniques (HRS) and with linear scattering techniques (dynamic light scattering or quasi-elastic light scattering). But most interestingly, the HRS intensity and depolarization results for these purple membrane patches could be analyzed toward a threefold symmetrical, partly octupolar superstructure of individual protein molecules, in complete agreement with independent structural studies (Hendrickx et al. 1996). The first hyperpolarizability value for the individual protein molecule could be traced to the retinal moiety embedded in the protein (Hendrickx et al. 1995). Studies of the model compounds indicate that the retinal is covalently linked to the protein backbone as the protonated retinal Schiff base.

Based on the known threefold symmetry for the bacteriorhodopsin protein molecule in the purple membrane patches, a relation between the individual β_{zzz} for retinal and the β_{ZZZ} for the patches was determined as a function of the angle θ between the plane of the purple membrane (perpendicular to the Z-axis) and the polyene chain (along the molecular z-axis of retinal). The relation for the experimental depolarization ratio as a function of this angle θ could be derived and is displayed in Figure 2.10.

From the experimentally observed depolarization ratio of $\rho = 0.56$, a value for the angle between the PM plane and the retinal chromophore of $(10 \pm 1)°$ could be obtained. Note from the figure how the limiting values for the pure dipolar structure (angle $\theta = 90°$ with respect to the PM plane, ρ equal to 0.2) and for the pure D_{3h} octupolar structure (angle $\theta = 0°$ with respect to the PM plane, ρ equal to 2/3) are nicely consistent with the proposed model. The experimental value of 0.56 indicates large octupolar

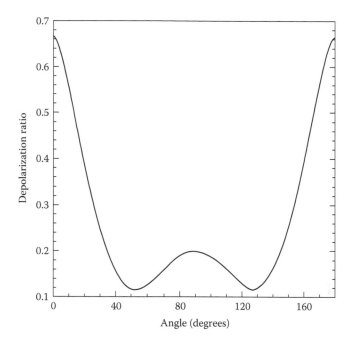

Figure 2.10 Depolarization ratio as a function of tilt angle out of plane for retinal chromophore in purple membrane patches.

and minor dipolar contributions to the second-order nonlinear optical response of bacteriorhodopsin when present in the purple membrane.

Recently, the second-order nonlinear optical properties of the photoswitchable fluorescent protein Dronpa have been investigated (Asselberghs et al., n.d.). A significantly different value for the dipolar first hyperpolarizability β_{zzz} has been found for the emissive and for the dark state. However, the depolarization ratios for both states, emissive and dark, were found to be identical. Of course, for the emissive state, this depolarization ratio had to be determined as a function of amplitude modulation frequency to arrive at the high-frequency limit, which is not influenced by the fluorescence. Fortunately, for strong emissive states with relatively long fluorescence lifetimes, the complete demodulation can be achieved within the bandwidth of our instrument and accurate depolarization ratios could be obtained. The identical values for the depolarization ratio of emissive and dark states show that the difference in optical properties (emissive and also second-order nonlinear optical) are not due to a change in conformation of the chromophore in the fluorescent protein.

A particularly nice example of solution-phase structure determination has been obtained for the important example of calixarenes (Hennrich et al. 2005). Calixarenes are supramolecular structures built from a number of repeat units that are covalently linked in a circular fashion to form a chalice-like structure. Unilateral substitution ("wide-rim" or "narrow-rim" substitution) leads to noncentrosymmetric structures that, when substituted with strong electron-donating or withdrawing groups, can exhibit large second-order nonlinearities. Of course, these depend on the opening angle of the chalice, and on the number of substituents. For an even number of substituents with assumed single major hyperpolarizability tensor component β_{zzz} and the angle 2θ between the individual components, the resulting supramolecular β_{ZZZ} along the supramolecular dipolar Z-axis can be written as $\beta_{ZZZ} = n\beta_{zzz} (\cos \theta)^3$. From the determination of this β_{ZZZ} and the β_{zzz} of the relevant model compound it has been possible to determine the average opening angle for a set of nine calix[4] arenes in solution and to compare these with a number of angles from x-ray diffraction from solid samples (Hennrich et al. 2007). Clearly, for an opening angle 2θ between the chromophores of 90°, maximum addition of the nonlinear response is obtained resulting in $\beta_{ZZZ} = n\beta_{zzz}$, whereas for an opening angle of 180° and for an even number of substituents, a centrosymmetric planar structure would be obtained with $\beta_{ZZZ} = 0$. Note how an odd number of substituents would result in an octupolar structure, with nonzero first hyperpolarizability, even in the planar conformation with opening angle 180°.

Note that for the solution-phase structure determination of the calixarenes, the HRS depolarization ratio was not used. In some cases, the dipolar nature of the model compounds and the calixarene supramolecular structure could be confirmed by the value for this ratio being close to 0.2. In other cases, this value was affected by multiphoton fluorescence and was, hence, significantly higher. Note that the HRS intensities were also affected by the fluorescence, but that the hyperpolarizability values themselves were corrected for this systematic artifact, as has been explained in Section 2.2.

A final example of solution-phase structure determination, this time necessarily with the use of the HRS depolarization ratio, is in the field of so-called coupled oscillators (Uyeda et al. 2002). These are also covalently linked individual chromophores, but the term seems to be reserved for the combination of large chromophores with extensive individual oscillator strength, such as the polypyridyl metal complexes covalently linked to large metalloporphyrin structures. The latter provides large polarizability while the former not only adds to it, but also induces the noncentrosymmetry, necessary for the second-order nonlinear response. Dialkylaminobenzene donor and nitrobenzene or nitrothiophene acceptor groups, meso-linked by ethynyl bridges to a central Zn(II) porphyrin, show the expected yet very large dipolar response. Replacing the thermally less stable dialkylamino group with the more stable polypyridyl Ru(II) moiety leads to a slightly higher value for the first hyperpolarizability but with an unusual dispersion.

The most exciting observation, however, was the very large second-order nonlinear optical response for the symmetrically meso-substituted porphyrin rings (Duncan et al. 2008). The true nature of this even-order response was carefully checked with fluorescence demodulation techniques. The symmetry of the structures was confirmed to be noncentrosymmetric octupolar by the low HRS depolarization ratio. As already explained in Section 2.3, this symmetry elucidation was only possible by the combination of accurate HRS intensity measurements (no odd-order fluorescence contributions to the second-order nonlinear response from noncentrosymmetric structures only), depolarization measurements, and chemical topology, but without the need for crystallization. Hence, the structural information is highly relevant, since obtained in solution and at room temperature.

With the examples shown, we wanted to demonstrate the power of the second-order nonlinear optical molecular scattering technique that HRS is for the solution-phase structure determination. HRS is a versatile tool, not only for the screening of the first hyperpolarizability of molecules but also for understanding structural chemistry. However, great experimental care needs to be taken to avoid contamination with odd-order optical responses, which are not limited to noncentrosymmetric structures. We have provided the experimental details to discriminate second-order HRS from odd-order one- and two-photon fluorescence and have shown with some well-chosen systems the applicability and importance of HRS for solution-phase structure elucidation.

References

Asselberghs, I., K. Clays, A. Persoons, A.M. McDonagh, M.D. Ward, and J.A. McCleverty, *Chem. Phys. Lett.*, 368 (2003), 408.

Asselberghs, I., C. Flors, L. Ferrighi, E. Botek, B. Champagne, H. Mizuno, R. Ando, A. Miyawaki, J. Hofkens, M. Van der Auweraer, and K. Clays. Manuscript in preparation.

Asselberghs, I., G. Hennrich and K. Clays, *J. Phys. Chem. A*, 110 (2006), 6271.

Asselberghs, I., Y. Zhao, K. Clays, A. Persoons, A. Comito, and Y. Rubin, *Chem. Phys. Lett.*, 364 (2002), 279.

Atkins, P.W., *Physical Chemistry*, Oxford, UK: Oxford University Press, 1978.

Aubert, V., V. Guerchais, E. Ishow, K. Hoang-Thi, I. Ledoux, K. Nakatani, and H. Le Bozec, *Angew. Chem. Int. Ed.*, 47 (2008), 577.

Birge, R.R., and C.-F. Zhang, *J. Chem. Phys.*, 92 (1990), 7178.

Botek, E., M. Spassova, B. Champagne, I. Asselberghs, A. Persoons, and K. Clays, *Chem. Phys. Lett.*, 412 (2005), 274.

Botek, E., M. Spassova, B. Champagne, I. Asselberghs, A. Persoons, and K. Clays, *Chem. Phys. Lett.*, 417 (2006), 282.

Boubekeur-Lecaque, L., B.J. Coe, K. Clays, S. Foerier, T. Verbiest, and I. Asselberghs, *J. Am. Chem. Soc.*, 130 (2008), 3286.

Boutton, C., K. Clays, A. Persoons, T. Wada, and H. Sasabe, *Chem. Phys. Lett.*, 286 (1998), 101.

Champagne, B., and D.M. Bishop, *Adv. Chem. Phys.*, 126 (2003), 41.

Clays, K., E. Hendrickx, M. Triest, T. Verbiest, A. Persoons, C. Dehu, and J.L. Brédas, *Science*, 262 (1993), 1419.

Clays, K., and A. Persoons, *Phys. Rev. Lett.*, 66 (1991), 2980.

Clays, K., and A. Persoons, *Rev. Sci. Instrum.*, 63 (1992), 3285.

Clays, K., A. Persoons, and L. De Maeyer, *Adv. Chem. Phys.*, 85 (1994), 455.

Clays, K., K. Wostyn, K. Binnemans, and A. Persoons, *Rev. Sci. Instrum.*, 72 (2001a), 3215.

Clays, K., K. Wostyn, K. Binnemans, and A. Persoons, *J. Phys. Chem. B*, 105 (2001b), 5169.

Clays, K., K. Wostyn, A. Persoons, S. Maiorana, A. Papagni, C.A. Daul, and V. Weber, *Chem. Phys. Lett.*, 372 (2003), 438.

Coe, B.J., *Chem. Eur. J.*, 5 (1999), 2464.

Coe, B.J., *Acc. Chem. Res.*, 39 (2006), 383.

Coe, B.J., J.L. Harries, L.A. Jones, I. Asselberghs, K. Clays, B.S. Brunschwig, J.A. Harris, J. Garín, and J. Orduna, *J. Am. Chem. Soc.*, 128 (2006), 12192.

Coe, B.J., J.A. Harris, L.J. Harrington, J.C. Jeffery, L.H. Rees, S. Houbrechts, and A. Persoons, *Inorg. Chem.*, 37 (1998), 3391.

Coe, B.J., S. Houbrechts, I. Asselberghs, and A. Persoons, *Angew. Chem. Int. Ed.*, 38 (1999), 366.

Cyvin, S.J., J.E. Rauch, and J.C. Decius, *J. Chem. Phys.*, 43 (1965), 4083.

Delaire, J.A., and K. Nakatani, *Chem. Rev.*, 100 (2000), 1817.

Duncan, T.V., K. Song, S.-T. Hung, I.M.A. Nayak, A. Persoons, T. Verbiest, M.J. Therien, and K. Clays, *Angew. Chem. Int. Ed.*, 47 (2008), 2978.

Flipse, M.C., R. de Jonge, R.H. Woudenberg, A.W. Marsman, C.A. Van Walree, and L.W. Jenneskens, *Chem. Phys. Lett.*, 245 (1995), 297.

Giraud, M., A. Léaustic, R. Guillot, P. Yu, P.G. Lacroix, K. Nakatani, R. Pansu, and F. Maurel, *J. Mater. Chem.*, 17 (2007), 4414.

Heesink, G.J.T., A.G.T. Ruiter, N. van Hulst, and B. Bolger, *Phys. Rev. Lett.*, 71 (1993), 999.

Hendrickx, E., C. Boutton, K. Clays, A. Persoons, S. van Es, T. Biemans, and E.W. Meijer, *Chem. Phys. Lett.*, 270 (1997), 241.

Hendrickx, E., K. Clays, A. Persoons, C. Dehu, and J.L. Brédas, *J. Am. Chem. Soc.*, 117 (1995), 3547.

Hendrickx, E., A. Vinckier, K. Clays, and A. Persoons, *J. Phys. Chem.*, 100 (1996), 19672.

Hennrich, G., M.T. Murillo, P. Pados, K. Song, I. Asselberghs, K. Clays, A. Persoons, J. Benet-Buchholz, and J. de Mendoza, *Chem. Comm.*, 21 (2005), 2747.

Hennrich, G., M.T. Murillo, P. Prados, H. Al-Saraierh, P.E. Georghiou, D.W. Thompson, I. Asselberghs, and K. Clays, *Chem. Eur. J.*, 13 (2007), 7753.

Hennrich, G., A. Omenat, I. Asselberghs, S. Foerier, K. Clays, T. Verbiest, J.L. Serrano, *Angew. Chem. Int. Ed.*, 45 (2006), 4203.

Kaatz, P., Shelton, D.P., Molecular physics, 88 (1996), 683.

Lamère, J.F., P.G. Lacroix, N. Farfan, J.M. Rivera, R. Santillan, and K. Nakatani, *J. Mater. Chem.*, 16 (2006), 2913.

Malaun, M., R. Kowallick, A.M. McDonagh, M. Marcaccio, R.L. Paul, I. Asselberghs, K. Clays, A. Persoons, B. Bildstein, C. Fiorini, J.-M. Nunzi, M.D. Ward, and J.A. McCleverty, *J. Chem. Soc.*, Dalton Transactions, 20 (2001), 3025.

Malaun, M., Z.R. Reeves, R.L. Paul, J.C. Jeffery, J.A. McCleverty, M.D. Ward, I. Asselberghs, K. Clays, and A. Persoons, *Chem. Comm.*, 01 (2001), 49.

Mançois, F., L. Sanguinet, J.L. Pozzo, M. Guillaume, B. Champagne, V. Rodriguez, F. Adamietz, L. Ducasse, and F. Castet, *J. Phys. Chem. B*, 111 (2007), 9795.

Mançois, F., V. Rodriguez, J.L. Pozzo, B. Champagne, and F. Castet, *Chem. Phys. Lett.*, 427 (2006), 153.

Marder, S.R., C.B. Gorman, F. Meyers, J.W. Perry, G. Bourhill, J.L. Brédas, and B.M. Pierce, *Science*, 265 (1994), 632.

Moreaux, L., O. Sandre, M. Blanchard-Desce, and J. Mertz, *Opt. Lett.*, 25 (2000), 320.

Morrison, I.D., R.G. Denning, W.M. Laidlaw, and M.A. Stammers, *Rev. Sci. Instrum.*, 67 (1996), 1445.

Noordman, O.F.J., and N.F. van Hulst, *Chem. Phys. Lett.*, 253 (1996), 145.

Olbrechts, G., R. Strobbe, K. Clays, and A. Persoons, *Rev. Sci. Instrum.*, 69 (1998), 2233.

Oliva, M.M., J. Casado, J.T. Lopez Navarrete, G. Hennrich, M.C. Ruiz Delgado, and J. Orduna, *J. Phys. Chem. C*, 111 (2007), 18778.

Prasad, P.N., and D.N. Williams, *Introduction to Nonlinear Optical Effects in Molecules and Polymers*, New York: Wiley, 1991.

Sanguinet, L., J.L. Pozzo, M. Guillaume, B. Champagne, F. Castet, L. Ducasse, E. Maury, J. Soulié, F. Mançois, F. Adamietz, and V. Rodriguez, *J. Phys. Chem. B*, 110 (2006), 10672.

Sanguinet, L., J.L. Pozzo, V. Rodriguez, F. Adamietz, F. Castet, L. Ducasse, and B. Champagne, *J. Phys. Chem. B*, 109 (2005), 11139.

Shin, S., and M. Ishigame, *J. Chem. Phys.*, 89 (1988), 1892.

Song, N.W., T.-I. Kang, S.C. Jeoung, S.-J. Jeon, B.R. Cho, and D. Kim, *Chem. Phys. Lett.*, 261 (1996), 307.

Terhune, R.W., P.D. Maker, and C.M. Savage, *Phys. Rev. Lett.*, 14 (1965), 681.

Uyeda, H.T., Y. Zhao, K. Wostyn, I. Asselberghs, K. Clays, A. Persoons, and M. Therien, *J. Am. Chem. Soc.*, 124 (2002), 13806.

Verbiest, T., K. Clays, A. Persoons, F. Meyers, and J.L. Brédas, *Opt. Lett.*, 18 (1993), 525.

Verbiest, T., K. Clays, C. Samyn, J. Wolff, D. Reinhoudt, and A. Persoons, *J. Am. Chem. Soc.*, 116 (1994), 9320.

chapter three

Characterization of interfaces, surfaces, and thin films

3.1 Second-harmonic generation and sum-frequency generation from surfaces: General principles

3.1.1 Introduction

The penetration depth of optical radiation into condensed matter is large, in general, which makes the isolation of an interface contribution difficult. However, deeper understanding of the underlying physics of the optical response, combined with advances in instrumentation, has allowed the contribution from the interface to be identified. Particularly important has been the recognition that symmetry differences between the bulk and interface can be exploited, as can interface electronic and vibrational resonances. The potential of optical techniques for interfaces, surfaces, and thin films characterization, with submonolayer resolution, began to be recognized in the 1980s. Thus, in the field of surface science, linear optical techniques are usually flexible, but it has been demonstrated among the literature that they sometimes suffer from the lack of surface specificity and sensitivity. Besides, in the field of nonlinear optics (NLO), second-harmonic generation (SHG) and sum-frequency generation (SFG) have developed into powerful tools for characterizing thin films and surfaces. SHG experiments are generally simple to perform, typically yield sub-monolayer interface-specific sensitivity with minimal optimization, and can be done with fairly inexpensive equipment. Coherent SHG and SFG are dipole forbidden in randomly oriented media, but allowed at the interface between two such media where inversion symmetry is broken. As a result, SHG can often be used to probe surfaces and interfaces in situ under normal conditions with negligible contributions from the bulk. Thus, surface-specific studies have been carried at several interfaces (liquid–liquid, liquid–solid, liquid–air, air–solid). Also, measurements of molecular order and orientation at surfaces and interfaces can yield a unique and powerful description of surface systems down to a monolayer.

As mentioned in Chapter 1, the lowest order nonlinear optical response of materials produces three-wave mixing phenomena, which includes SHG and SFG, and these phenomena may be surface sensitive at nondestructive power densities. For centrosymmetric media, a simple calculation shows that the surface effect should be at least comparable in size to the higher-order nonlocal bulk effects (Shen 1984; Guyot-Sionnest et al. 1986). However, experimentally, one still needs to be able to unambiguously separate the surface and bulk contributions and usually one considers symmetry arguments to interpret the experimental data. In the most general second-order nonlinear response of a system, three-wave mixing occurs when two incident fields of frequency (ω_1 and ω_2) combine to produce a third field of frequency (ω_3), where $\omega_3 = \omega_1 \pm \omega_2$.

As we saw in Chapter 1 (Section 1.3), different combinations and degeneracies of the input fields can produce SHG, optical rectification, SFG, and difference-frequency generation (DFG; Kuzyk and Dirk 1998). It is important to note that these processes are coherent and the radiating fields have a well-defined direction. For example, SHG from a surface in the usual reflection geometry in vacuum emerges along the path of the reflected primary beam, which dramatically simplifies the detection (Figure 3.1). These coherent properties, in contrast with hyper-Rayleigh scattering (also called harmonic light scattering; see Chapter 2) which is an incoherent process, enable SHG and SFG to be used as surface probes, even though the cross-section for the process is typically four orders of magnitude lower than the Raman scattering cross-section (which also is mainly an incoherent inelastic scattering process).

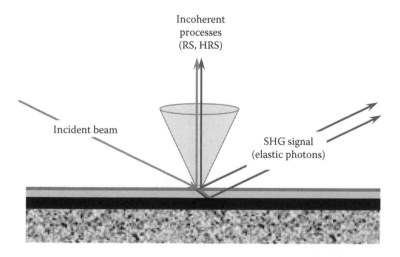

Incoherent processes (RS, HRS)

Incident beam

SHG signal (elastic photons)

Figure 3.1 Reflected coherent SHG and incoherent (Rayleigh, hyper-Rayleigh light) scattering processes.

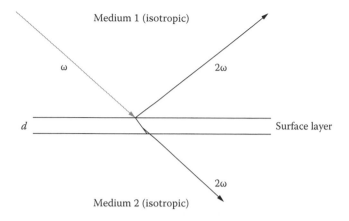

Figure 3.2 Sketch of SHG from an interfacial film between two isotropic media.

Beyond this, including local and nonlocal contributions, the induced second-order polarization from an interfacial layer of thickness d bounded by two bulk isotropic media (Figure 3.2) can be written as (in Gaussian units)

$$P_i(\omega_3) = \chi^{eee}_{ijk} E_j(\omega_1) E_l(\omega_2) + \chi^{eeq}_{ijkl} E_j(\omega_1) \nabla_k E_l(\omega_2) \tag{3.1}$$

Beside the usual (local contribution) electric dipolar susceptibility χ^{eee}_{ijk}, we have introduced a nonlocal one, χ^{eeq}_{ijkl}, which is of the electric-quadrupole character and takes the electric gradient field contribution into account (see Section 1.6). Higher-order contributions are neglected here. Surface SHG can arise from the higher-order term mentioned in Equation 3.1 due to the large field gradients normal to the surface, but it can also arise from the electric-dipole term because the inversion symmetry is broken at the surface due to a polar ordering of the interfacial molecules (Eisenthal 1996).

As mentioned by Shen (1984), as far as the interface layer signal is concerned, two approaches can be taken. One uses the solutions for a three-layer system; the middle layer corresponding to the surface layer has a thickness (d) approaching zero. This is the classical approach that is used to describe SHG/SFG in bulk media, such as crystals, polymers, or *Langmuir–Blodgett* (LB) films. The second approach, which is often used in the field of surface science, assumes no clear boundary between the surface layer and medium 2 (the second layer, usually a substrate) and it uses a combined nonlinear polarization induced in the substrate and the surface layer:

$$P_{eff}(\omega_3) = P_{medium2}(\omega_3) + P_{surface}(\omega_3) \tag{3.2}$$

where $P_{medium2}(\omega_3)$ is described by Equation 3.1 and $P_{surface}(\omega_3)$ is function of a surface susceptibility $\chi_s^{(2)}$ given by

$$P_{i,surface}(\omega_3) = \sum \chi_{S,ijk}^{(2)} E_j(\omega_1) E_k(\omega_2)$$ (3.3)

As we saw in Chapter 1, the second-order nonlinear susceptibility for a layer with a finite thickness, $\chi_{medium}^{(2)}$, or an interface/surface, $\chi_S^{(2)}$, is a polar tensor of rank 3.

Note also that in SFG or SHG, the polarization state of the radiated frequency is dependent on the $\chi^{(2)}$ tensor components, the polarization states of the two incident beams, the linear optical properties of the sample, and the surrounding media and optical path, that is, the incidence angles of the two incoming fundamental beams. To illustrate this, let us consider the case of SHG from α-quartz (D_3, uniaxial symmetry), which is one of the reference materials used in SHG. Assuming Kleinman symmetry is valid, the only nonzero element is d_{11} ($d_{14} = 0$)* and the polarization components are:

$$P_x^{(2)} = d_{11}\left(E_x^2 - E_y^2\right)$$

$$P_y^{(2)} = -2d_{11}E_x E_y$$ (3.4)

$$P_z^{(2)} = 0$$

As depicted in Figure 3.3, let θ be the angle between the propagation vector and the crystalline axis (the optic axis) that lies in the plane of incidence XZ, and ϕ the angle between the electric field (linear polarization) and the plane of incidence. Note that the x-axis together with the z-axis of the crystal lay in the horizontal plane of incidence XZ. For an arbitrary linearly polarized input field E_ω, the relation with the field components in the xyz coordinate system is given by a simple coordinate transformation:

$$\begin{pmatrix} E_x \\ E_y \\ E_z \end{pmatrix} = \begin{pmatrix} \cos\theta\cos\phi \\ \sin\phi \\ \sin\theta\cos\phi \end{pmatrix} E_\omega$$ (3.5)

---

* This hypothesis is valid since d_{14} scales only ~1% of d_{11} at 1064 nm because of the weak dispersion of quartz in the visible/near-infrared range.

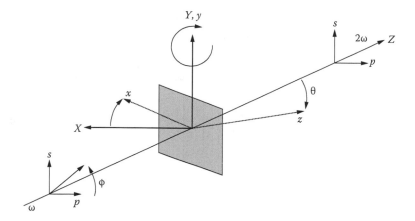

Figure 3.3 Scheme of a polarized SHG experiment with an α-quartz plate (xyz frame) at an incident angle θ.

The combination of Equations 3.4 and 3.5 gives the following effective susceptibilities:

$$P_x^{(2)} = d_{11}(\cos^2\theta\cos^2\phi - \sin^2\phi)E_\omega^2$$

$$P_y^{(2)} = -2d_{11}(\cos\theta\cos\phi\sin\phi)E_\omega^2 \tag{3.6}$$

$$P_z^{(2)} = 0$$

One can now clearly see that the polarization is dependent on the incidence angle, susceptibility component, and polarization angle.

Therefore, for a p-polarized input (in the plane of incidence XZ) where $\phi = 0°$, these relations reduce to:

$$P_x^{(2)} = \left(d_{11}\cos^2\theta\right)E_\omega^2$$

$$P_y^{(2)} = 0 \tag{3.7}$$

$$P_z^{(2)} = 0$$

and it is clear that the second-harmonic field will only have a p-polarized component.

For a s-polarized input (out of the plane of incidence) where $\phi = 90°$, we obtain:

$$P_x^{(2)} = -d_{11}E_\omega^2$$

$$P_y^{(2)} = 0 \tag{3.8}$$

$$P_z^{(2)} = 0$$

and also in this case there will only be a p-polarized second-harmonic component.

3.1.2   The wave equation for SFG and SHG

We aim here to give a simple physical picture of how these frequency components are generated. As a general case, let's consider sum-frequency generation as reported in Figure 3.4.

The general equation of propagation of the electric field $E(r,t)$ is

$$\nabla \times (\nabla \times E(r,t)) + \frac{1}{c^2}\frac{\partial^2}{\partial t^2}D(r,t) = 0 \tag{3.9}$$

It is often convenient to decompose the electric displacement D into its linear ($D^{(1)}$) and nonlinear ($D^{(2)}$) parts as

$$D = D^{(1)} + D^{(2)} \tag{3.10}$$

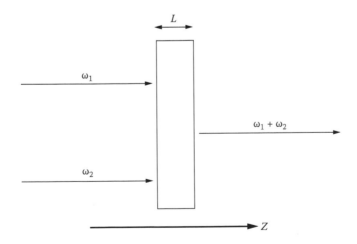

Figure 3.4 Sum-frequency generation in a nonlinear plate, with thickness L, at normal incidence, along the Z direction.

From the constitutive relations, we can express $D^{(1)}$ in terms of $P^{(1)}$, the part of the polarization $P(r,t)$ that depends linearly on the electric field, as well as in terms of the dielectric tensor ε:

$$D^{(1)} = E + 4\pi P^{(1)} = (1 + 4\pi \chi^{(1)})E = \varepsilon E \tag{3.11}$$

The nonlinear displacement $D^{(2)}$ can be expressed as a function of $P^{(2)}$, the nonlinear polarization that has been detailed previously in Chapter 1. It reads

$$D^{(2)} = 4\pi P^{(2)} \tag{3.12}$$

Now taking Equations 3.10 through 3.12 into Equation 3.9, we obtain the general wave equation of the electric field $E(r, t)$

$$\nabla \times (\nabla \times E(r,t)) + \frac{1}{c^2} \frac{\partial^2}{\partial t^2} D^{(1)}(r,t) = -\frac{4\pi}{c^2} \frac{\partial^2}{\partial t^2} P^{(2)}(r,t) \tag{3.13}$$

This equation has the form of a driven (inhomogeneous) wave equation in which the nonlinear medium acts as a source term that appears in the right-hand side of this equation. In a homogeneous medium that has no second-order response, the nonlinear source term is zero and Equation 3.13 admits solutions of the form of free waves propagating with velocity c/n, where $n = \sqrt{\varepsilon}$ is the linear optical constant of the medium. In the case of a lossless material, n is the linear index of refraction of the medium.

Since we consider waves propagating along the z direction as depicted in Figure 3.4, $E(r,t)$ and $P(r,t)$ at frequency ω may be expressed as

$$E(r,t) = [E_0(r,t)\exp[i(kz - \omega t)] + c.c.] \tag{3.14}$$

$$P(r,t) = [P_0(r,t)\exp[i(kz - \omega t)] + c.c.] \tag{3.15}$$

where *c.c.* means complex conjugate and $k = n\omega/c$ is the wave vector of the field in the medium. In the stationary regime, which involves monochromatic continuous-wave input, we have $E_0(r,t) = E_0(r)$ and $P_0(r,t) = P_0(r)$. This condition is valid even when using pulsed laser sources down to the picosecond regime.

Under certain conditions, the general form of the wave Equation 3.13 can be simplified if we consider the following identity from vector calculus:

$$\nabla \times (\nabla \times E(r,t)) = \nabla(\nabla.E(r,t)) - \Delta E(r,t) \tag{3.16}$$

For an isotropic medium far from any discontinuity, the Maxwell equation $\nabla.D = 0$ implies $\nabla.E = 0$ and the wave equation can be simplified in the form

$$\Delta E(r,t) - \frac{1}{c^2}\frac{\partial^2}{\partial t^2}D^{(1)}(r,t) = \frac{4\pi}{c^2}\frac{\partial^2}{\partial t^2}P^{(2)}(r,t) \qquad (3.17)$$

More generally, in linear and nonlinear optics, the term $\nabla(\nabla.E(r,t))$ from Equation 3.16 is small, even when it does not vanish and especially when the slowly varying amplitude approximation is valid.* Following Equation 3.14, the Laplacian of the electric field reads

$$\Delta E(r,t) = \left\{ \left[\left(\frac{\partial^2}{\partial z^2} + 2ik\frac{\partial}{\partial z} - k^2 \right)E_0(r,t) + \left(\frac{\partial^2}{\partial x^2} + \frac{\partial^2}{\partial y^2} \right)E_0(r,t) \right]\exp[i(kz-\omega t)] + c.c. \right\}$$

$$(3.18)$$

Assuming the slowly varying amplitude approximation we have then

$$\frac{\partial^2 E_0(r,t)}{\partial z^2} \ll 2ik\frac{\partial E_0(r,t)}{\partial z} \qquad (3.19)$$

Therefore, the Laplacian of the electric field $E(r, t)$ can be expressed as

$$\Delta = \left[\left(2ik\frac{\partial}{\partial z} - k^2 \right) + \Delta_\perp \right] \qquad (3.20)$$

where $\Delta_\perp = \left(\frac{\partial^2}{\partial x^2} + \frac{\partial^2}{\partial y^2} \right)$ is a term responsible for diffraction effects.

Now the question is to derive the specific nonlinear optical interactions from the wave equation that we have established. Let us consider the simple case of SFG in a lossless nonlinear material involving a monochromatic transverse, infinite plane wave (stationary regime). We assume the configuration described in Figure 3.4 where the applied fields interact with the medium surface at normal incidence. In the following sections, we will consider situations that are more general. Under this assumption,

---

* The slowly varying amplitude approximation is valid when the amplitude of the field $A(r, t)$ varies slowly over distances of the order of the wavelength $\lambda = 2\pi n/k$ and during times of the order of the optical period $T = 2\pi/\omega$.

we have $E_0(r,t) = E_0(z)$. Thus, the Laplacian of the electric field is simplified since no diffraction effects occur $(\Delta_\perp = 0)$ and the left-hand side of the wave Equation 3.17 simplifies to

$$\left[2ik\frac{\partial}{\partial z} - k^2\right]E(z) - \frac{\varepsilon}{c^2}\frac{\partial^2}{\partial t^2}E(z) = \left[2ik\frac{\partial}{\partial z} - k^2\right]E(z) + \frac{\omega^2\varepsilon}{c^2}E(z) = \left[2ik\frac{\partial}{\partial z}\right]E(z)$$

$$(3.21)$$

Finally, the wave equation is expressed as

$$2ik\frac{\partial E(z)}{\partial z} = \frac{4\pi}{c^2}\frac{\partial^2}{\partial t^2}P^{(2)}(z) \tag{3.22}$$

The wave equation must hold for each frequency component of the electric field, particularly at the sum-frequency ω_3. We can represent the nonlinear source term that radiates at $\omega_3 = \omega_1 + \omega_2$ as

$$P_3^{(2)}(z) = 2\chi^{(2)}(z)E_1(z)E_2(z) \tag{3.23}$$

where the subscript 3 refers to the sum-frequency ω_3.
Also the applied fields can be written as

$$E_i(z) = [E_{i,0}(z)\exp[i(k_iz - \omega_it)] + c.c.] \tag{3.24}$$

with subscript $i = 1,2,3$ referring to the three different frequencies.
The nonlinear polarization reads

$$P_3^{(2)}(z) = 2\chi^{(2)}(z)E_{1,0}(z)E_{2,0}(z)\exp\{i[(k_1 + k_2)z - (\omega_1 + \omega_2)t]\} + c.c$$

$$= P_{3,0}^{(2)}(z)\exp\{i[(k_1 + k_2)z - (\omega_1 + \omega_2)t]\} + c.c$$

$$= P_{3,0}^{(2)}(z)\exp\{i[(k_1 + k_2)z - (\omega_3)t]\} + c.c \tag{3.25}$$

The right-hand part of Equation 3.22 may be rewritten as

$$\frac{4\pi}{c^2}\frac{\partial^2}{\partial t^2}P_3^{(2)}(z) = -\frac{4\pi\omega_3^2}{c^2}P_3^{(2)}(z) \tag{3.26}$$

Finally, the wave equation for the ω_3 frequency of the electric field is

$$2ik_3 \frac{\partial E_3(z)}{\partial z} = -\frac{4\pi\omega_3^2}{c^2} P_3^{(2)}(z) \tag{3.27}$$

We can cancel the factor $\exp[-i\omega_3 t]$ on each side (see Equations 3.24 and 3.25) and write the resulting equation as

$$\frac{\partial E_{3,0}(z)}{\partial z} = i \frac{4\pi^2}{n_3 \lambda_3} P_{3,0}^{(2)}(z) \exp[i\Delta k.z] \tag{3.28}$$

where $\Delta k = (k_1 + k_2) - k_3$ is called the wave-vector mismatch.

Equation 3.28 is known as the coupled-amplitude equation because it shows how the amplitude of the ω_3 wave varies because of its coupling to the two other (input) waves. Similar coupled-amplitude equations exist for ω_1 (as a function of ω_2 and ω_3) and ω_2 (as a function of ω_1 and ω_3). However, usually in SFG or SHG the amplitudes of the two input fields are assumed to be constant (no pumps depletion, that is, weak interaction regime), which corresponds to the approximation that only a negligible part of energy from the input fields is converted into the SHG or SFG frequency. Therefore, under this assumption, only the coupled amplitude (Equation 3.28) of the SFG or SHG signal is considered. When the nonlinear medium is homogeneous, the amplitude of the nonlinear polarization is usually constant, $P_{3,0}^{(2)}(z) = P_{3,0}^{(2)}$, which understands that the nonlinear susceptibility is constant, $\chi^{(2)}(z) = \chi^{(2)}$. Additionally, in case of phase matching, when $\Delta k = (k_1 + k_2) - k_3 = 0$, the amplitude $P_{3,0}^{(2)}(z)$ of the SFG wave increases drastically with z. This condition, known as the condition of phase matching, indicates that the nonlinear polarization and the generated waves maintain a fixed phase relation that efficiently converts the incident power. From a microscopic point of view, this corresponds to a proper phasing of the individual dipoles and their amplitude adds in the forward direction. Hence, the total power radiated by the dipoles is proportional to the square of the number of scatterers that participate.

In the general case, when no phase matching occurs, in the weak interaction regime the amplitude of the SFG signal is low and at the exit of the NLO medium it is obtained by integrating Equation 3.28 from $z = 0$ to $z = L$:

$$E_{3,0}(L) = i \frac{4\pi^2}{n_3 \lambda_3} P_{3,0}^{(2)} \int_0^L \exp[i\Delta k.z] dz = i \frac{4\pi^2}{n_3 \lambda_3} P_{3,0}^{(2)} \left[\frac{\exp[i\Delta k.L] - 1}{i\Delta k} \right] \tag{3.29}$$

Using Equation 3.23 we obtain:

$$E_{3,0}(L) = i\frac{8\pi^2}{n_3\lambda_3}\chi^{(2)}E_{1,0}E_{2,0}\left[\frac{\exp[i\Delta k.L]-1}{i\Delta k}\right] \tag{3.30}$$

Since the intensity of the three waves is given by $I_i = n_i c/2\pi\,|E_{i,0}(L)|^2$ (i = 1, 2, or 3), after simplification the sum-frequency intensity reads in terms of the input intensities:

$$I_3 = \frac{32\pi^5}{n_1 n_2 n_3 c\lambda_3^2} I_1 I_2 [\chi^{(2)}]^2 \left|\frac{\exp[i\Delta k.L]-1}{i\Delta k}\right|^2 \tag{3.31}$$

The squared modulus in this equation can be reworked as

$$
\begin{aligned}
\left|\frac{\exp[i\Delta k.L]-1}{i\Delta k}\right|^2 &= \left(\frac{\exp[i\Delta k.L]-1}{\Delta k}\right)\left(\frac{\exp[-i\Delta k.L]-1}{\Delta k}\right) \\
&= \left(\frac{\cos\Delta kL + i\sin\Delta kL - 1}{\Delta k}\right)\left(\frac{\cos\Delta kL - i\sin\Delta kL - 1}{\Delta k}\right) \\
&= \left(\frac{2-2\cos\Delta kL}{\Delta k^2}\right) = \left(\frac{2\sin^2(\Delta kL/2)}{\Delta k^2}\right) = L^2\frac{\sin^2(\Delta kL/2)}{(\Delta kL/2)^2} \\
&= L^2\sin c^2(\Delta kL/2)
\end{aligned}
\tag{3.32}
$$

Hence, the intensity becomes

$$I_3 = \frac{32\pi^5}{n_1 n_2 n_3 c\lambda_3^2} I_1 I_2 [\chi^{(2)}]^2 L^2 \sin c^2(\Delta kL/2) \tag{3.33}$$

A few remarks should be noted from Equation 3.33:

1. The SFG intensity is proportional to $I_3 = [\chi^{(2)}]^2/n_1 n_2 n_3$. It is necessary to have a strong NLO medium to gain a SFG (or SHG) signal.
2. $I_3/I_1 \propto I_2$ (or $I_3/I_2 \propto I_1$). It is necessary to use a laser with high peak power without inducing dielectric breakdown of the medium.
3. The SFG intensity is proportional to $L^2\sin c^2(\frac{\Delta k.L}{2}) = L^2\sin c^2(\frac{\pi}{2}\frac{L}{L_c})$. The oscillatory behavior of the intensity is connected to the coherence length $L_c = \pi/\Delta k$, which is the coherent buildup length of the

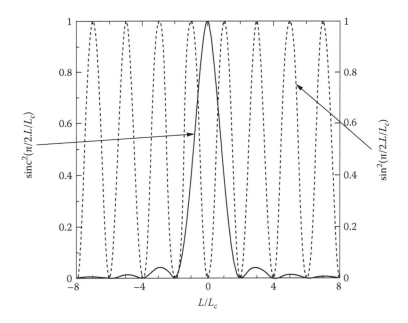

Figure 3.5 Effect of phase mismatch on the efficiency of SFG. Every $(2n + 1)L/L_c$ period, a maximum occurs but with a decrease of the oscillation amplitude as the phase mismatch (n) increases. Every $(2n)L/L_c$ period a minimum occurs, corresponding to an output out of phase and power flows back from the ω_3 wave into the ω_1 and ω_2 waves.

interaction (Figure 3.5). The wave vector mismatch is entirely included in this factor and the mixing process decreases as $\Delta k.L$ increases. Notably, for a very thin NLO medium ($L \ll L_c$ and $L \rightarrow 0$), the sin c function is close to 1 and the SFG intensity is proportional to the square of the layer thickness L.

Equation 3.33 predicts a dramatic decrease of the efficiency of SFG when the phase matching condition $\Delta k = 0$ is not satisfied. In the general case, the phase matching condition is often difficult to achieve and Maker et al. first observed this oscillatory behavior in 1962 (Figure 3.6). They observed interference patterns, called Maker fringes, by focusing the output of a pulsed ruby laser into a single crystal of quartz and measuring the output SHG intensity at different angles of incidence. Thus, the rotation of the sample introduces a variation of the effective path length (L) through the plate.

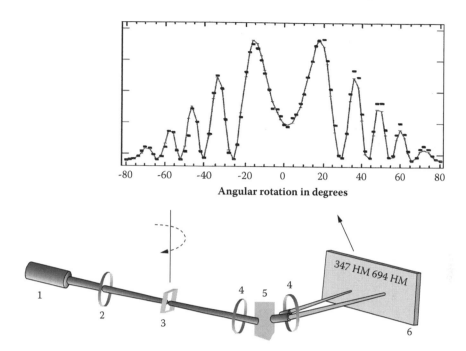

Angular rotation in degrees

Figure 3.6 Scheme of the first experimental observation of second-harmonic generation (Maker et al. 1962). 1, ruby laser; 2, focalizing lens; 3, quartz plate sample; 4, collimator lenses; 5, prism; 6, photograph plate.

3.2 *Experimental techniques and equipment*

3.2.1 *Absolute and relative measurements*

For more than three decades, several experimental techniques have been widely used for determining second-order nonlinear optical (NLO) coefficients $\chi_{ijk}^{(2)}$ and, among them, there are two methods that determine absolute values of the NLC coefficients. The first method is SHG phase matching and the second is parametric fluorescence, which gives good accuracy (Shoji et al. 1997). These absolute methods are difficult to implement since an accurate determination of the fundamental and radiated beam powers is required. As a result, these experiments are rarely used.

Two other relative methods make a good alternative and are typically preferred: the Maker fringe technique and the wedge technique. In both techniques, a pulsed laser is made incident on the sample, while the

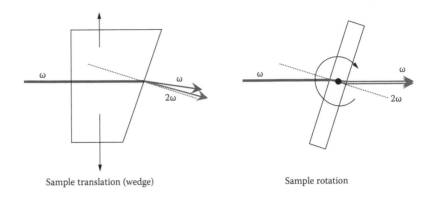

Sample translation (wedge)   Sample rotation

Figure 3.7 Variation of the path length in the wedge technique (left) and in the Maker fringe technique (right).

intensity of the SFG- or SHG-radiated light is monitored as a function of the effective path length. The optical path varies with the sample translation in the case of the wedge technique, whereas it varies with the sample rotation in the Maker fringe technique (Figure 3.7).

The wedge technique has been widely described (Chemla and Kupecek 1971). However, one of its main disadvantages is that the sample must be made in the shape of a wedge and this requirement is not always easy to achieve. On the other hand, since multiple reflections from fundamental and harmonic beams are eliminated, excluding the occurrence of fringe interferences due to the modification of the path length with the translation of the sample, the data treatment is quite easy and straightforward.

3.2.2   Maker fringe technique

The Maker fringe experimental method is based on the observation of SHG from NLO materials, not phase matchable in the used frequency range, relative to a standard reference. Standard reference materials such as α-quartz or ammonium dihydrogen phosphate (ADP) plates are often used to obtain absolute values of the NLO components, even though their reference values have considerably changed over the last 15 years (Kuzyk and Dirk 1998; Bosshard et al. 2000). The samples are usually in the shape of planar structures (crystal plates, layers deposited on a substrate, etc.) or planar interfaces (liquid-air interfaces, LB films, etc.), which make this experiment fairly simple and widely applicable.

Since the second harmonic response of a medium is proportional to the square of the incident light intensity, SHG and SFG measurements normally employ high-powered pulsed laser systems. Pulsed laser systems, with wavelengths ranging between 500 nm and 2500 nm, are the

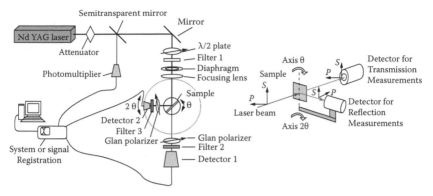

Figure 3.8 SHG Maker fringe experimental apparatus for the detection in transmission or reflection. A high-powered pulsed laser (Nd:YAG operating here at 1064 nm) is used as a light source. With a semitransparent mirror, a small portion of the laser beam is sent to a reference channel that also includes laser fluctuations. The fundamental beam is linearly polarized, filtered, and focused onto the sample. Typical angles of incidences range between −80° and 80° from the surface normal. The second harmonic beam, either collected in transmission or in reflection, (θ - 2θ coupling) is separated from the fundamental beam by a selective density filter and a band-pass interferential filter (532 nm). A rotating analyzer (GLAN) selects either p (in plane) or s (out of plane) polarized SHG signals that are detected by a photomultiplier tube.

main excitation sources for three-wave mixing experiments in materials. A typical experiment might use nanosecond pulses of a few microjoules to millijoules of energy and with an energy density kept below 1 J/cm² to avoid any laser-induced damage effects. Q-switched Nd:YAG lasers have been widely used at 1064 nm excitation and frequency doubled at 532 nm; whereas Ti:sapphire and optical parametric oscillator (OPO) systems are used for wavelength-dependent studies. A scheme of the typical experimental apparatus used in SHG Maker fringe is shown in Figure 3.8.

Glan–Taylor prisms, together with a half-wave plate or double Fresnel rhomb, are used for polarization selection. It is very important to remove the SHG signal generated in these optical components in the input with optical filters. The SHG signal is weak, and either a gated integrator or gated photon counter is used. Conversion efficiencies are usually very low, on the order of 10^{-9}% or lower. A small piece of the fundamental beam is often deviated to a reference channel that monitors either the fundamental power (as is exemplified in Figure 3.8) or the SHG signal from a reference standard material. The output of the reference channel normalizes the SHG signal from the sample and removes the effect of any fluctuations in the laser power during the experiment. SHG experiments can be performed either in transmission or in reflection, but the former only will work for transparent or weak absorbing materials. Typical angles of

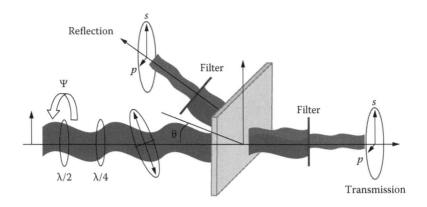

Figure 3.9 Scheme of a surface SHG experiment in transmission or in reflection with a continuous polarization scan. A combination of a rotating half-wave plate (with angle ψ) and a fixed quarter-wave plate (vertical fast axis) address all possible polarizations, from linear to elliptical and circular polarization.

incidence range from normal incidence to 60°–80° for transmission and from quasi-normal incidence to 60°–80° for reflection.

The sensitivity and selectivity of the SHG technique may become less effective when studying thin films, surfaces, or interfaces, with the traditional Maker fringe technique as previously described. Thus in surface SHG experiments, continuous polarization scans of the input (sometimes output) beam at fixed incidence angle are preferred (as is depicted in Figure 3.9).

The fundamental beam, initially polarized out of the plane of incidence (s), is passed through a combination of a rotating half-wave plate and a fixed quarter-wave plate (vertical fast axis) to address all possible polarizations, from linear to elliptical and circular polarization. The incident beam is focused on the sample at a fixed incident angle (frequently at 45° in the literature). Transmitted (or reflected) second-harmonic light is resolved into components polarized parallel (p) and perpendicular (s) to the plane of incidence (frequently horizontal). In Figure 3.9, the second-harmonic intensity is recorded versus the rotation angle of the half-wave plate (angle ψ). An alternative setup can be used where only a quarter-wave plate is involved. The rotation of the quarter-wave plate rotation addresses all states of polarization (linear, circular, and elliptical), but the sequences of the polarization states are slightly different (Figure 3.10).

Thus, when input and output beam polarizations and geometry are properly chosen, the various elements of $\chi^{(2)}_{medium}$ (or $\chi^{(2)}_S$; see Equation 3.3) can be selectively measured by SHG. The symmetry of $\chi^{(2)}_{medium}$ reflects the symmetry of the average molecular arrangement at the medium or surface.

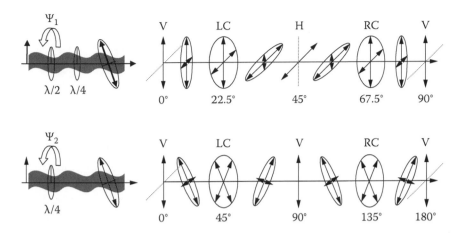

Figure 3.10 Optical schemes and corresponding states of polarization used for continuous polarization scans. Linear vertical (V), linear horizontal (H), left circular (LC), and right circular (RC) states of polarizations are indicated. Intermediate left or right elliptical polarizations are not reported.

SHG is well adapted to probe the structural symmetry of a thin film or a monolayer adsorbed at an interface. In the latter case, when studying a horizontal liquid interface, another setup that requires a polarization scan at a fixed incident angle is necessary, but with a vertical plane of incidence (see, e.g., Eisenthal 1996). A possible scheme of the experimental setup to observe a liquid/monolayer interface is depicted in Figure 3.11. In that case, it is interesting to introduce an additional rotating $\lambda/4$ plate (or a rotating $\lambda/2$ plate and a fixed $\lambda/4$ plate) after the polarizer to address

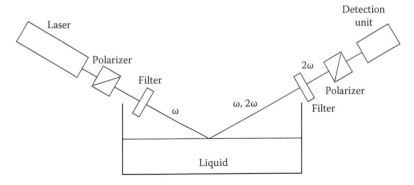

Figure 3.11 Scheme of apparatus to observe monolayers at a liquid interface (vertical plane of incidence).

elliptical and circular polarization states of the incident wave, as discussed earlier.

3.3   Probing the properties of interfaces, surfaces, and thin films

In the following sections, we will focus on the interpretation of SHG patterns and we will illustrate the application of SHG to the study of interfaces, surfaces, and thin films. We will first give examples of transparent thin oxide glass films followed by a few illustrations in the domain of organic LB or polymer thin films.

3.3.1   Interpreting Maker fringe patterns

To the best of our knowledge, as far as we consider bulk contributions, Herman and Hayden (1995) were the first to take into consideration the problem of anisotropic and absorbing materials and they proposed an analytical solution for isotropic absorbing materials in which no pump depletion occurs (weak interaction regime). In fact, it is noteworthy to recall that many dedicated and restrictive theories for analyzing optical SHG in reflection or transmission, using linearly (vertical s-, or horizontal p-) polarized fundamental beams, birefringent and/or absorbing media, multilayers, and so forth have been introduced (Bloembergen and Pershan 1962; Jerphagnon and Kurtz 1970; Kuzyk et al. 1989; Herman and Hayden 1995). In particular, several authors have proposed a 4 × 4 matrix formulation for Maker fringe analyses in biaxially birefringent materials (Bethune 1989, 1991; Braun et al. 1997, 1998). As first introduced for linear processes (Abéles 1972; Berreman 1972) and then developed by Bethune for nonlinear ones (Bethune 1989, 1991), the problem of optical harmonic generation and mixing in multilayer anisotropic media can be treated by using an optical transfer matrix approach.

More recently, a general theoretical formalism designed to quantify linear optical and second-order nonlinear optical responses of achiral or chiral anisotropic materials in planar structures has been introduced (Rodriguez and Sourisseau 2002; Rodriguez 2008). In particular, this theory includes linear optical activity that is governed by the gyrotropic components and second-harmonic generation optical rotatory dispersion, the magnitude of which depends on the ratio of chiral and achiral $\chi^{(2)}$-components.

One typical example is dihydrogen phosphate (KDP), which exhibits an optically active $\overline{4}2m$ quadratic symmetry. This class of symmetry is a special case since there is one independent off-diagonal gyrotropic component, $\rho_{xy} = -\rho_{yx}$ that induces optical rotation of light along [110] directions. In this case, the gyrotropic tensor is traceless, which means that both enantiomer forms are present in the structure. The non-vanishing

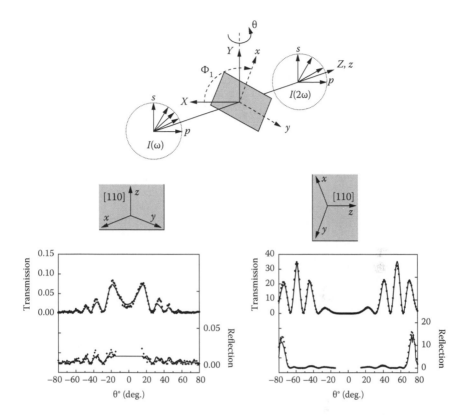

Figure 3.12 Example of reflected and transmitted SHG signals (at 1064 nm) of a [110] KDP plate. Experimental (dotted lines) and calculated (solid lines) 45°-*p* signals with vertical *c*-axis (left) and horizontal *c*-axis (right). Absolute intensities are in arbitrary units.

SHG components of symmetry class $\overline{4}2m$ are *xyz* and *zxy*.* Typical experimental and calculated (beyond our simple considerations) (SHG patterns obtained in reflection and transmission) of KDP single-crystals are shown in Figure 3.12 and corresponding results are given in Table 3.1 (see Rodriguez 2008 for further details).

3.3.2   Oxide photonic glasses

Homogeneous glasses are natural centrosymmetric materials, which implies the even nonlinear dielectric susceptibility $\chi^{(2)}$ is zero. Thus, from symmetry considerations, there is no nonlinear optical response of the

---

* *xyz* and *zxy* are Kleinman disallow components here (*xyz* ≠ *zxy*) because of the strong optical dispersion of the material.

Table 3.1 Results Obtained (at 1064 nm) with a [110] KDP
Plate (Class of Symmetry −42 m)

Thickness = 328.49 μm	1064 nm	532 nm
n_x	1.4938	1.5087
n_z	1.4592	1.4692
$\Delta_n = n_z - n_x$	−0.0345	−0.0394
ρ_{xy} (°/mm)	+2.4	+20.8
Tensor $\chi^{(2)}$ (pm/V)	$xyz = +0.374$	
	$zxy = +0.354$	

Note: Accuracies are: index of refraction ±10⁻⁴, gyrotropic component ±0.2°, all $\chi^{(2)}$ components ±2 10⁻³ pm/V.

second order in any homogeneous glassy material. However, second-order
nonlinear (SONL) responses in glasses may be observed ($\chi^{(2)} \neq 0$) that result
from the action of an intensive laser irradiation (Sasaki and Ohmori 1981),
an external electric field (thermal poling; Myers et al. 1991), or the presence
of inhomogeneities in the glass. Inhomogeneous glasses may be obtained
by means of nanostructuring (Sigaev et al. 2002) or by the crystallization of
phases (Kao et al. 1994) with large optical nonlinearity of the second order.
The technique of thermal poling has been widely used to induce SHG
responses in oxides glasses. The method consists of applying a DC electric
field at a high temperature close to the T_g of the material before cooling
the sample while the direct current field is kept on. The second-harmonic
generation Maker fringe technique is currently used to characterize the
induced second-order properties of thermally poled glasses.

3.3.2.1   Induction of SONL properties in oxide glasses

The origin of SONL responses in glasses as a result of the action of an
external electric field was reported for the first time by Myers et al. in
1991. The experimental procedure is rather simple. A fused silica glass
plate of 1.6 mm thickness was set between a silicon wafer electrode and
a steel electrode before heating up to 523–598 K. Then an electrical voltage of 500–5000 V was applied during 15 minutes before switching off
the heating to cool the material to room temperature but with the DC
field on (Figure 3.13). As a result, SHG interference patterns were observed
when varying the incident angle of the incident beam light that come
mostly from a thin layer, with a thickness of several microns, at the anode
interface. The best $\chi^{(2)}$ value from this SONL layer in silica-based glass,
determined by the optical Maker fringe method, is ~1 pm/V, which is
approximately 1/30 of the LiNbO₃ largest component $\chi^{(2)}_{zzz}$ ~32 pm/V
(Dmitriev et al. 1997). In addition, these pioneering authors observed a

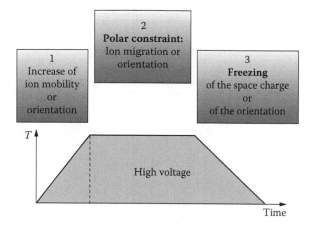

Figure 3.13 Principle of a poling experiment: First, heating increases the mobility of the charges or dipolar moieties. Second, a high voltage, at a constant high temperature, is applied, which induces either charges migration (creation of a volume space charge) or the reorientation of dipolar moieties along the field gradient. Third, keeping the DC field on maintains the polar constraint, while cooling down to room temperature to freeze the space charge or dipolar orientation.

linear dependence of the SHG signal with the applied voltage. A comparison of SHG signals obtained with the thermal poling of silica glasses under various impurities content of Na⁺, ranging from 1 to 20 ppm, has shown that the SONL signal was directly correlated with the impurity content. It could be concluded that SONL properties in silica glasses are connected to the migration of mobile charge carriers, particularly Na⁺ cations. This ionic migration in a thin layer is at the origin of a volume space charge that induces a static electric field embedded in the glass at the anode surface. The observation of SHG due to the presence of the static electric field is referred to as electric field induced second-harmonic generation or EFISH. Strictly speaking, the effect is a third-order nonlinear optical effect, described by a fourth-rank tensor $\chi^{(3)}$ that can be related to second-order susceptibility $\chi^{(2)}$ (see Equation 3.34).

SONL properties induced by thermal poling (or EFISH) have been extensively studied in silica and other inorganic glasses. Kazansky and Russell (1994) reported that SONL responses in poled glasses result from a "frozen" electric field at the near-anode surface of the glass that reads:

$$\chi^{(2)}_{EFISH} = 3\chi^{(3)}E_{stat}$$

(3.34)

where E_{stat} is the frozen electric field strength that ranges from 0.1 to 1.0 GV/m, just below the electric breakdown in a silica glass, for instance.

The local symmetry of a SONL optical layer in a poled glass belongs to point-group $C_{\infty v}$, which is a polar uniaxial group oriented along the applied DC field, perpendicular to the film surface. Consequently, assuming Kleinman's symmetry, which is realistic in usual glasses since they are transparent and weakly dispersive, SONL properties of poled glasses may be described by two independent components: $\chi_{zzz}^{(2)}$ and $\chi_{zxx}^{(2)} = \chi_{xxz}^{(2)}$. In addition, assuming the EFISH mechanism described by Equation 3.34, the symmetry requirement for the third-order nonlinear dielectric susceptibility tensor in an isotropic material (i.e., a glass; Butcher and Cotter 1990) implies one additional restriction for the resulting tensor $\chi^{(2)}$ submitted to a DC electric field:

$$\chi_{zxx}^{(2)} = \frac{1}{3}\chi_{zzz}^{(2)} \tag{3.35}$$

The application of a DC electric field to a glass, at a sufficient temperature that favors certain ion mobility, triggers mostly the migration of mobile cations Na^+ and K^+ to a less extent, whereas anionic moieties are known to be less mobile. In fact, the induction of an embedded DC electric field based on such relative charge mobility has been previously observed in ionic crystals (Von Hippel et al. 1953). Apart from that model of charge carriers migration, another possible model that accounts for reorientation and/or structural reordering of local dipoles has been proposed (Mukherjee et al. 1994) for the description of the nonlinear optical glasses, in analogy with poled organic polymers (Burland et al. 1994):

$$\chi_{Orientation}^{(2)} \propto \frac{N.\beta.\mu_0.E_{dc}}{5k_B T} \tag{3.36}$$

where β is the second-order polarizability of the local dipole, μ_0 is the electrical permanent dipole moment, N is the density number of dipoles, k_B is the Boltzmann constant, and T is the temperature. This model, originally applied to poled organic polymers dominated by van der Waals interactions, exhibits a first-order linear dependence of the SONL response with the DC electric field and has also been used for silica glasses.

Figure 3.14 sketches the origin of the embedded electric DC field in a thermally poled glass containing mobile positive carriers, Na^+, whereas negative carriers are bounded by the glass structure. The anode is a blocking electrode and the nonblocking cathode makes possible any ionic migration through the electrode; sodium cations are then free to move out from the cathode interface of the glass. Therefore, a depletion of mobile cations at the anode interface would favor the creation of a negative space-charge layer where negative charge carriers are bounded. The occurrence

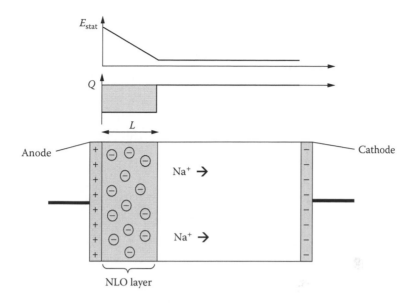

Figure 3.14 Model of migration of the cations and of the "frozen" electric field during thermal poling in a silica glass. Electric field (E_{stat}) and charges (Q) distributions in the sample are reported at the top of the figure.

of such a layer during the poling process would result in a partial screening of the external applied electric DC field in the volume of the glass.

It has been demonstrated that whatever the silica glasses are made from (either from a quartz-melt using a gas-jet flame or from a sol-gel synthesis), impurities of sodium cations or mobile protons at an amount of several parts per million are sufficient to ensure the screening of a rather large external field (Imai et al. 1998). Element spatial distribution studies in poled silica glasses have shown a partial migration of several mobile carriers (cations K^+, Na^+, Li^+, H^+) as a result of the action of a DC electric field. Since all these charge carriers were at the level of impurities in the glass (Alley et al. 1999), the elemental analyses were performed by a method of a secondary ionic mass spectrometry. However, even at the level of a few ppm, these mobiles cations have been moved away from the anode glass interface and the width of the SONL layer could be increased with time during the poling process. $\chi^{(2)}$ values of poled silicate glasses Suprasil™ and Optosil™ are 0.10 and 0.02 pm/V, respectively (Quiquempois et al. 2000). It is obvious here that the SONL property of silicate glasses depends strongly on the content of mobile charge carriers. Additionally, SHG features of Suprasil poled glasses indicate that SONL properties unusually involve a larger layer probably spreading the entire glass sample. The dependency of $\chi^{(2)}$ and the width of the SONL layer, when modifying the concentration of charges in silica glasses and with time, have been studied using either the Maker fringe technique (Kazanky et al. 1993; Qui et al. 2001) or by a stepwise etching of the surface with hydrogen fluoride combined with a periodic recording of SHG-signals (Henry et al. 1996; Kudlinski et al. 2005a, 2005b). These studies have shown that although the $\chi^{(2)}$ signal remains unchanged for months, the NLO layer width does not vary within 4 to 15 days after poling; 6 months later it decreases approximately 30%. Following Quiquempois et al. (2002), the greatest possible value of a nonlinear dielectric susceptibility of the second order $\chi^{(2)}$ in a silica glass is sized up. According to the value of the third-order susceptibility $\chi^{(3)}_{zzzz} = 2 \cdot 10^{-22}$ m^2/V^2 in silicate glasses and the breakdown potential of thin films (which is estimated around $E \sim 8 \cdot 10^8$ V/m), the maximum theoretical value $\chi^{(2)}_{zzz}$ is 0.5 pm/V. Therefore, to derive higher SONL responses, it is necessary to study the thermal poling processes and SONL properties of glasses with other compositions.

3.3.2.2   Thickness and profile effect of the NLO layer in poled oxide glasses

The Maker fringe technique is well adapted to probe the SHG response of poled glasses. From the preceding paragraph, we may infer that the glass plate is decomposed in two layers: first, a bulk layer that is isotropic (i.e., without SHG activity) and, second, a nonlinear layer at the anode interface with a thickness of a few microns (Figure 3.15).

In the thin NLO layer, the SHG tensor exhibits a $C_{\infty v}$ symmetry (uniaxial polar). Furthermore, since the material is transparent in the visible range and assuming a weak optical dispersion, we can apply Kleinman relations that gives $d_{15} = d_{31}$. Hence, the tensor is given by:

$$d_{C_{\infty v}} = \begin{pmatrix} 0 & 0 & 0 & 0 & d_{15}=d_{31} & 0 \\ 0 & 0 & 0 & d_{15}=d_{31} & 0 & 0 \\ d_{31} & d_{31} & d_{33} & 0 & 0 & 0 \end{pmatrix} \qquad (3.37)$$

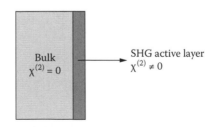

Figure 3.15 Representation of the two layers in a poled bulk glass plate.

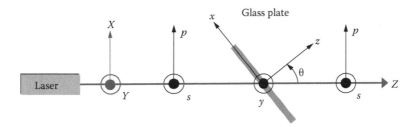

Figure 3.16 Scheme of the Maker fringe experiments where *XYZ* is the laboratory frame and *xyz* is the internal glass plate frame. The sample is rotated along the *Y* (//*y*) vertical direction with an angle θ, and *XZ* is the plane of incidence. The input and output polarizations are either linearly *p*-polarized (in plane) or *s*-polarized (out of plane).

The d_{33}/d_{31} ratio depends upon the origin of the induced SHG by the poling process. On the one hand, if the NLO signal arises from any dipole orientation (or a structural modification), as given in Equation 3.36, the stronger the orientation, the higher the ratio. On the other hand, if the signal is induced by the interaction of an internal static field with the third-order susceptibility, $\chi^{(3)}$, following Equations 3.34 and 3.35, the ratio is $d_{33}/d_{31} = 3$. So far, only the latter effect (EFISH) has been evidenced. Hence, Maker fringe simulations include at least two parameters: the thickness of the layer and a constant value of the d_{33} coefficient, while d_{31} is constrained to $d_{33}/3$. Let us define *XYZ* as the laboratory frame and *xyz* as the internal frame of the glass plate where the optical axis is along the *z*-direction given by the static electric field (Figure 3.16).

The NLO polarization in the internal *xyz* frame reads:

$$P_x = 2d_{31}E_xE_z \qquad\qquad (p-polarized\ output)$$

$$P_y = 2d_{31}E_yE_z \qquad\qquad (s-polarized\ output) \qquad (3.38)$$

$$P_z = d_{31}\left(E_x^2 + E_y^2\right) + d_{33}E_z^2 \qquad\qquad (p-polarized\ output)$$

If we consider only p-($E_x \neq 0$, $E_z \neq 0$, $E_y = 0$) or s-($E_y \neq 0$, $E_x = E_z = 0$) polarized input light, it can be seen that no NLO polarization comes out of the plane of incidence (output *s*-polarized), that is, along the *y* (//*Y*) direction. So whatever the incidence angle θ is, any incident *s*- or *p*-polarized electric field will only give rise to NLO polarization in the *xz* plane of incidence that corresponds to a *p*-polarized output. Hence, *p*–*p* and *s*–*p* polarization configurations are currently performed in poled glasses and poled materials, if the poling has been achieved along the *z*-direction. However, note that the SHG study can be largely improved by also combining the

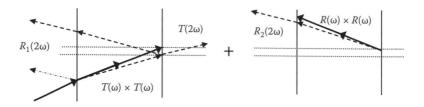

Figure 3.17 Schematic of the two SHG reflected types of signal $R_1(2\omega)$ (from fundamental forward waves) and $R_2(2\omega)$ (from fundamental backward waves) that can strongly interfere. Note that the transmitted SHG signal $T(2\omega)$ is only due to the forward incident waves.

polarization scan technique at fixed angles of incidence (as described in Section 3.2), but it is not yet currently adopted by the community.

Typically in the case of poled glasses, one of the main difficulties is to determine the thickness of the NLO layer. A possible answer, using the Maker fringe technique, is to consider both the transmitted and reflected harmonic signals to gain information on the layer. In the case of thin films where the thickness is below or close to the coherence length (L_c), the optical path $L(\theta)$ does not vary sufficiently, even at large incidence. Consequently, no oscillatory pattern occurs and the patterns obtained in transmission are less indicative of the thickness of the NLO film.* As mentioned in Section 3.1.2, for a thin NLO layer with thickness l, the transmitted SHG signal scales as L^2. In that case, typically, a maximum occurs around 60 degrees in the pp transmitted scheme (55 degrees in the sp configuration) because the Fresnel factors decrease to zero at larger incidence. In contrast, the reflected harmonic signal is more sensitive to thin layers, as it is also the case in linear optics (see, e.g., Abéles 1972; Azzam and Bashara 1977). Note that in the reflection mode, there are two types of harmonic waves that interfere (Figure 3.17): one type comes from the reflection of the SHG signal ($R_1(2\omega)$) from the forward incident waves ($T(\omega)$), whereas the other type comes from the reflected incident waves ($R(\omega)$) that generate a backward SHG signal ($R_2(2\omega)$). Several examples of possible resulting interferences can be found elsewhere (Rodriguez 2008; Rodriguez and Sourisscau 2002).

As a demonstrating example of this in the field of photonic glasses, let us focus on the induced second-order optical nonlinearity generated in thermally poled sodium borophosphate niobium glasses of compositions $(1 - x)[0.95\text{Na(PO}_3) + 0.05\ \text{Na}_2\text{B}_4\text{O}_7] + x\ \text{Nb}_2\text{O}_5$. Here, the choice of the composition (x = 0.35, 0.45) has been oriented by previous evidence of a large increase of the macroscopic $\chi^{(3)}$ values with the ratio of niobium

---

* The equivalent difficulty occurs in linear optics (ellipsometry measurements on a thin film, etc.): the product phase term $k.L = nL\omega/c$ is usually determined; a large set of (n, L) solutions is mathematically possible.

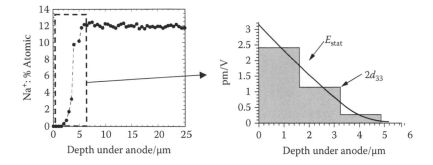

Figure 3.18 Left: Depth profiles of the atomic concentration of sodium from $0\ \mu m$ (anode surface) to a few micrometers inside the bulk glass. Right: Simulated stepwise profile SHG profile ($2d_{33}$ coefficient) obtained from the best fit to SHG Maker fringes. Following Equation 3.34, a comparison of the $2d_{33}$ stepwise profile is given together with the rescaled E_{stat} profile obtained from the Na$^+$ space charge profile reported on the left.

oxide (Cardinal et al. 1996, 1997). It has been also demonstrated that the electric field-induced SHG process (EFISH) is the predominant one in such glasses (see Equation 3.34). In particular, a space-charge migration of Na$^+$ at the anode side of the poled plates has been clearly evidenced (Figure 3.18). The depleted layer is characterized by a strong internal electric field $E_{stat}(z)$, decreasing almost linearly from the anodic surface ($z = 0$) to 3–4 μm deep inside the sample where it becomes zero (Dussauze et al. 2005). On the other hand, through an original analysis of both transmitted and reflected polarized Maker fringe patterns of the poled glass reported in Figure 3.19, it was possible to determine a linear $\chi^{(2)}(z)$ profile in very good agreement with the internal electric field profile $E_{stat}(z)$. In other words, for bulk glasses it was shown that the depth profile of $\chi^{(2)}(z)$ is correlated to the $\chi^{(3)}E_{int}(z)$ term (Equation 3.34).

3.3.3   Organic- and polymer-oriented materials

In the late 1980s and 1990s, so-called organic nonlinear optical materials captivated researchers in industry and academia. Nonlinear optical properties in organic and polymer systems are still under intensive study because, unlike inorganic systems where nonlinear optical phenomena arise from band structure or cooperative/dispersive effects, in organic and polymeric systems NLO effects originate in the virtual electron excitations occurring on the individual molecule or polymer chain units. Thus, interest in organic materials as sources of second-order nonlinear optical materials has centered on molecular materials in which the nonlinearity originate mainly from an appropriate collection of relatively

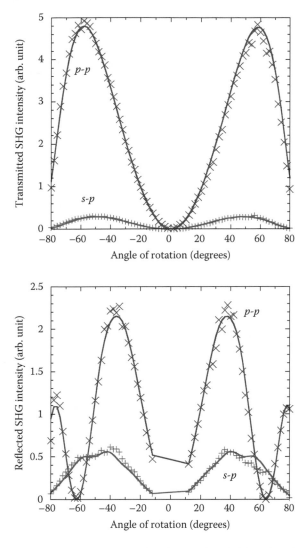

Figure 3.19 Experimental (crosses) and calculated (solid lines) polarized *pp* and *sp* Maker fringes obtained in transmission (top) and reflection (bottom) following a depth profile. Note in the *pp* reflected pattern (bottom) the nonlinear brewster angle around ±62° where no reflected *pp*-SHG intensity occurs.

independent molecular sources of optical nonlinearity. Most work has focused on the dipolar activity that arises from donor- and acceptor-substituted π-electron compounds possessing a permanent molecular dipole moment. A second point is the assembly of these chromophores in noncentrosymmetric arrangements in miniaturized nanoscale systems. Several approaches for generating polar macroscopic structures have

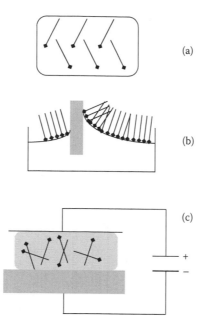

Figure 3.20 Examples of several methods for achieving NLO active thin films materials: (a) molecular crystals, (b) LB films, and (c) poled polymers.

been investigated (Figure 3.20). The earliest type that has been extensively studied was molecular crystals (Chemla and Zyss 1987). A second possibility is the fabrication of LB films. The LB technique provides a means for imposing a noncentrosymmetric structure onto a material that crystallizes in a centrosymmetric space group. Finally, a third possibility, the most widely used approach for the construction of micrometer-thick ordered organic thin films, involves electrical poling of organic polymer films that includes (doped or functionalized) nonlinear active moieties containing permanent electric dipoles.

3.3.3.1   Langmuir–blodgett (LB) films

In many cases, when the LB technique is used, noncentrosymmetric multilayers are made if transfer occurs when the substrate is being withdrawn (z-type) or immersed (x-type) through the floating monolayer. Y-type LB films, which are usually the more stable arrangement, are obtained by successive withdrawal of a polar substrate, resulting in successive head-to-head and tail-to-tail stacking of amphiphilic molecules that leads sometimes to an overall centrosymmetric structure. For a multilayer structure in which the noncentrosymmetry is preserved from layer to layer, the SHG efficiency is expected to increase quadratically with the number of layers (see Section 3.1.2).

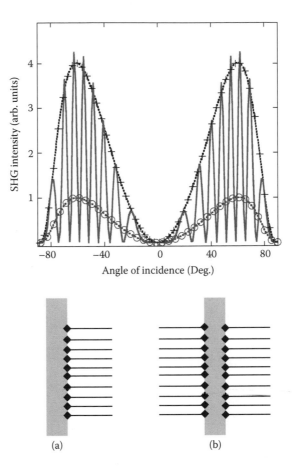

Figure 3.21 Typical SHG *pp* transmitted signals in *y*-type LB films. Dotted lines correspond to LB films with thickness *L* (circles) and 2*L* (crosses) coated on one side of the substrate (bottom: a). Full lines correspond to LB films with thickness *L* coated on both sides of the substrate (right: b).

In the case of vertical *y*-type deposition, a preferential alignment along the dipping direction is observed and strong nonlinear responses can be obtained. Also, typical SHG interference patterns are observed when each side of the substrate is coated, whereas the interference features disappear when only one side of the substrate is coated (Figure 3.21). This has been observed, for instance, in SHG investigations of LB films of 2-docosylamino-5-nitropyridine (DCANP; Dechter et al. 1988; Wijekoon et al. 1993; Provencher et al. 1995) and many other systems (Wijekoon et al. 1996; Yokoyama et al. 1998; Schrader et al. 1999; Schwartz et al. 2001). Thus, in diluted chromophores films, the square root of SHG intensity is generally scaled with the density number of molecules, $\sqrt{I^{2\omega}} \propto N$, as well as with the thickness of the NLO layer, $\sqrt{I^{2\omega}} \propto L$, as sketched in Figure 3.21. For both sides of coated samples, one observes that the number of interference fringes increases with the thickness of the substrate. As mentioned in the literature, this effect is due to the increase of the phase delay between the nonlinear responses of both layers and even a slight modification of the thickness of the substrate (a few micrometers) modulates the angular position of the minima and maxima. Additionally, in SHG measurements, since the nonlinear polarization is generated inside a NLO layer, the output harmonic energy flux is sensitive to the dielectric contrast between media that affects Fresnel factors and then introduces slight differences in the Maker fringe patterns. As a concluding remark, these trends are theoretically predicted for ideal coherent scattering layers. They apply down to the monolayer as long as it is able to radiate enough SHG coherent light, since, at this scale, local defects that trigger incoherent scattering probably occur.

3.3.3.2   Poled organic polymer films

For the design of materials suited for use in electro-optical devices, nonlinear optical chromophores incorporated in amorphous polymers are subjects of intense research activities since the early 1980s. Such materials can be produced in large-area devices and can be structured more easily in thin film device applications (see, e.g., Robinson et al. 1999). Interest in organic polymers as second-order NLO materials arises mainly from the fact that the nonlinearity is due to an appropriately arranged collection of relatively independent molecular sources of optical nonlinearity that molecular engineering can easily tune and optimize. Hence, major breakthroughs have been recently achieved in supramolecular organic electrooptic (EO) materials (Enami et al. 2007; Kim et al. 2007).

Most work has been focused on the dipolar nonlinearity, which arises from donor–acceptor substituted π-electron molecules, and the NLO chromophores are dispersed in or covalently attached to amorphous polymers (Figure 3.22). Additionally, enhancement by optical dispersion of the SHG

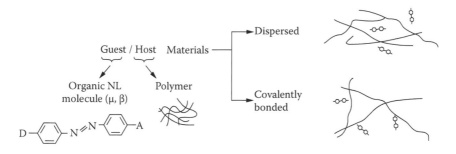

Figure 3.22 Guest-host materials are composed of NLO chromophores (A-π-D polar type) dispersed or grafted to the polymer backbone.

response (see Figure 1.6 in Chapter 1) is obtained when the charge-transfer absorption band is close to the harmonic excitation. We recall the reader that following Singer et al. (1987) the SHG optical response is proportional to the linear EO response (Pockels effect), since the NLO response has an exclusive electronic origin in organic materials.

In order to have finite second-order susceptibility, the polymer system must consist of chromophores oriented in such a way that the total system does not have a center of symmetry. Films of polymers are obtained usually by the drop- or spin-casting method using an adequate solvent that gives films good optical quality. However, such films are usually in-plane isotropic or even three-dimensional isotropic.

The polar ordering is accomplished by applying a DC electric field to the polymer at a temperature where the chromophores have a relative angular mobility. The principle of a poling experiment has been previously described (see Figure 3.13). A large number of approaches for generating polar materials from molecular systems have already been investigated, and the DC field poling of amorphous polymers is a largely developed technique (Singer et al. 1986; Lindsay and Singer 1995; Lee et al. 2000; Rodriguez et al. 2003). The thin polymer film can be sandwiched between two parallel conducting plates (interdigitated and coplanar electrodes) to induce a polar axis, but charge injection may also occur (Yitzchaik et al. 1991). However, high field strengths can be obtained in this contact poling method provided careful attention is given to material purity, dust-free thin film processing and electrode design.

Figure 3.23 gives a typical setup for corona poling, which is probably the most commonly used method. It is different from the contact poling one in the sense that shortcuts (occurring from residual dust or charged particles dispersed in the polymer, like Na^+) are avoided. The efficiency of this poling technique has been largely proved in the literature and a very large field exceeding 4 MV/cm can be obtained. Because of the extremely high electric field in the vicinity of the thin corona wire (or needle or grid),

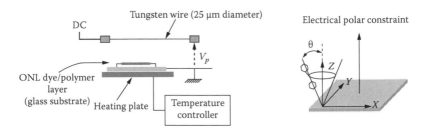

Figure 3.23 A typical setup for corona poling (left) that triggers a vertical alignment of the polar molecules along the electric field (right).

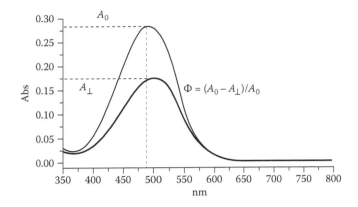

Figure 3.24 Typical ultraviolet-visible spectra of a poled film before (A_0) and after ($A_\perp$) poling.

electrons naturally present in the surrounding gas are accelerated to energies high enough to ionize molecules and atoms in the gas (usually air). Usually the polarity of the wire is positive (anode), and positive ions are swept out of the ionized region onto the film. An insulating top buffer layer (usually an acrylate-based polymer) is preferentially added to avoid any damage of the guest-host film during the poling treatment. After the poling, the top buffer layer is dissolved with water.

The anisotropy created by the application of a static electric field at a high temperature (Mortazavi et al. 1989) affects both the linear optical (LO) and the NLO properties.

On one hand, the anisotropy of the absorption coefficient can be used as a probe of the degree of orientation, $\langle \cos^2 \theta \rangle$, of the molecular polar groups (Figure 3.24). In a first approximation, one may describe the axial orientation of the chromophores by the order parameter Φ defined by $\Phi = 1 - A_\perp / A_0$ and proportional to $\langle \cos^2 \theta \rangle$, where A_0 and $A_\perp$ are respectively the absorbance maxima for the unpoled and poled samples at normal incidence. The order parameter Φ quantifies the linear dichroism (LD) of the poled film. Before poling, the maximum absorption A_0 corresponds to electric dipolar moments of transition randomly distributed. The more efficient the poling along the vertical axis, the more the dipolar chromophores are oriented along the vertical z-axis, that is, perpendicularly to the plane of the film. As a result, the absorption becomes anisotropic and it is maximum along the z-axis and minimum in the x–y plane of the film ($A_\perp$). Hence, for an unpoled sample, $\Phi = 0$ since $A_\perp = A_0$, and for an ideal perfectly poled sample, $A_\perp \rightarrow 0$ and $\Phi \rightarrow 1$. This order parameter definition holds valid under the assumption that, first, there is no degradation of the chromophores during the poling process and, second, the molecular ordering does not modify the overall characteristics of the absorption band. In fact, it is rather difficult to prove experimentally that the first condition is fulfilled. The second assumption may obviously be in contradiction with the large changes that could occur during an efficient poling process.

On the other hand, second-harmonic generation of light in a Maker fringe experiment is a standard method to quantify second-order nonlinearity $\chi^{(2)}$ and then the extent of the polar orientation, $\langle \cos\theta \rangle$ and $\langle \cos^3\theta \rangle$, of the molecular groups through the knowledge of their first-order hyperpolarizability, β.

3.4   Molecular orientation at surfaces

For the investigation of processes and properties related to interfaces, surfaces, and thin films, a wealth of particle-based techniques is available. However, different techniques are required for the investigation of molecular orientation, surface adsorption, and reactions at interfaces, as is proposed in the next paragraph. Among these, photon-based techniques play a crucial role. Besides linear optical techniques—such as ellipsometry, surface plasmon resonance, infrared spectroscopy, and surface-enhanced Raman spectroscopy—nonlinear methods such as SHG and SFG have been successfully applied to molecular layers. The combinations of intrinsic interface sensitivity, high versatility with respect to the choice of environment, and the possibility to determine molecular orientation by polarization-dependent measurements make SFG and SHG appealing to tackle problems related to molecular layers or molecular assemblies. However, the advantages of these techniques are contrasted with the main restriction of a sufficient NLO activity.

In this section, we will focus on molecular orientation determined by SHG since, without any loss of generality, it is the simplest and most popular second-order NLO technique to handle. First, we will define the relationship between the molecular and the bulk response and derive relations useful for the purpose. Then we will give further details and applications of molecular orientation at surfaces with uniaxial symmetry.

3.4.1   Relationship between bulk response
and molecular response

The general expression for the second harmonic intensity from a thin surface film has been given by Shen (1984):

$$I^{2\omega}(\omega) \propto |(\hat{e}^{2\omega}L^{2\omega}) \cdot \chi^{(2)} : (\hat{e}^{\omega}L^{\omega})(\hat{e}^{\omega}L^{\omega})|^2 \; I_\omega^2 \qquad (3.39)$$

where the terms $\hat{e}^{\Omega}$ are unit polarization vectors, L^{Ω} are appropriate Fresnel factors relating the external electric fields to the local fields in the film, and I_{ω} is the input intensity. The tensor $\chi^{(2)}$ is related to the (first) hyperpolarizability β tensor of individual chromophores within the film through orientational averages. Before going forward, let us briefly examine the underlying hypotheses. In the dipole approximation and for instantaneous response, we have seen in Chapter 1 that the polarization P, the bulk response, is defined as the dipole moment per unit volume and may be expressed as a power series in the external electric field (E):

$$P_I(t) = P_I^{(0)} + \chi_{IJ}^{(1)}(t)E_J(t) + \chi_{IJK}^{(2)}(t)E_J(t)E_K(t) + \cdots \tag{3.40}$$

where $P_I^{(0)}$ is the spontaneous polarization and the tensor quantity $\chi^{(n)}$ is the nth-order susceptibility. Similar to the bulk polarization in Equation 3.40, the molecular polarization (p) can be expanded in a power series in the local electric fields (F):

$$p_i(t) = \mu_{0,i} + \alpha_{ij}(t)F_j(t) + \beta_{ijk}(t)F_j(t)F_k(t) + \cdots \tag{3.41}$$

where $\mu_{0,i}$ is the molecular ground-state dipole moment, α_{ij} is the linear molecular polarizability, and β_{ijk} is the molecular hyperpolarizability.

In the dilute solute model (Figure 3.25) and under the hypothesis that the molecular units do not interact strongly, the macroscopic quantities can be related to the molecular ones by evaluating the thermodynamic average of the molecular polarization (p):

$$P_I(t) = N\langle p_i(t)^* \rangle_I \tag{3.42}$$

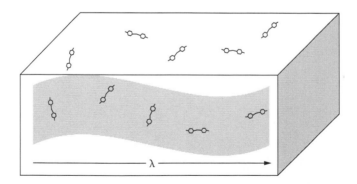

Figure 3.25 Representation corresponding to the dilute solute model where molecular units do not interact significantly. The size of the molecule is negligible compared to the wavelength λ, and the size of the volume where the average is performed corresponds to a large ensemble.

where N is the density number of molecules, $p(t)^*$ is the dressed molecular dipole (Kuzyk and Dirk 1998), that is, the effective internal moment, and where the brackets denote an orientational average.

$$\langle p_i(t)^* \rangle_I = \frac{\int \int \int d^3 \vec{\Omega} G(\vec{\Omega}) a_{Ii}(\vec{\Omega}) p_i(t)^*}{\int \int \int d^3 \vec{\Omega} G(\vec{\Omega})} \tag{3.43}$$

where $\vec{\Omega}$ represents the three Euler angles, $G(\vec{\Omega})$ is the orientational distribution function (ODF) that describes the orientations of the molecular moieties, and a_{Ii} is the rotation matrix that connects the dressed molecular frame (ijk) to the laboratory frame (IJK) given by (Figure 3.26):

$$a = \begin{pmatrix} (+\cos\theta\cos\phi\cos\psi - \sin\phi\sin\psi) & (+\cos\theta\sin\phi\cos\psi + \cos\phi\sin\psi) & (-\sin\theta\cos\psi) \\ (-\cos\theta\cos\phi\sin\psi - \sin\phi\cos\psi) & (-\cos\theta\sin\phi\sin\psi + \cos\phi\cos\psi) & (+\sin\theta\sin\psi) \\ (+\sin\theta\cos\phi) & (+\sin\theta\sin\phi) & (+\cos\theta) \end{pmatrix}$$

$$\tag{3.44}$$

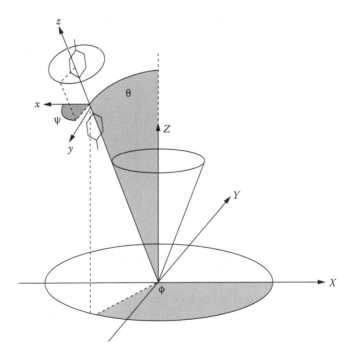

Figure 3.26 Schematic diagram of the molecular frame (xyz) and the laboratory frame (XYZ) with the corresponding Euler angles (ψ, θ, ϕ).

Because both the bulk dipole moment and the molecular polarization are expressed as a power series in the electric field, the nth-order term yields a relationship between the molecular and bulk susceptibilities:

$$P_I^{(0)} = N\langle \mu_{0,i}^* \rangle_I$$

$$\chi_{IJ}^{(1)} = N\langle \alpha_{ij}^* \rangle_{IJ} \qquad (3.45)$$

$$\chi_{IJK}^{(2)} = N\langle \beta_{ijk}^* \rangle_{IJK}$$

From Equation 3.45, the bulk tensor properties of the SHG second-order susceptibility can be calculated from the integral of the tensor properties of the molecule over all possible orientations but weighted by the distribution function as

$$\chi_{IJK}^{(2)} = N\left\langle \beta_{ijk}^* \right\rangle_{IJK} = N \int_0^{2\pi} d\phi \int_0^{2\pi} d\psi \int_{-1}^{+1} d(\cos\theta)\,\beta_{ijk}^*\, a_{Ii}\, a_{Jj}\, a_{Kk}\, G(\phi, \psi, \theta) \qquad (3.46)$$

with $\beta_{ijk}^* = \beta_{ijk}\, f_\omega^2 f_{2\omega}$, where f_Ω is the Lorentz–Lorenz correction factor introduced in Chapter 1 (see Equation 1.5), and $a_{\Delta\delta}$ is the rotation matrix element defined in Equation 3.44.

We have seen that the density of molecular units oriented in the direction is given by the orientational distribution function, G. For a surface in the presence of an electric field, E_P, or other orientational forces described by a mean field potential, U_T, the distribution function can be expressed as

$$G(\Omega) = \frac{\exp\left(-\frac{U_T}{k_B T}\right)}{\int d\Omega \exp\left(-\frac{U_T}{k_B T}\right)} \qquad (3.47)$$

where k_B is the Boltzmann constant. Thus, the aligning work to orientate the molecular units originates in both the short-range intermolecular interactions and the externally applied field, and it is comparable to $k_B T$. If we truncate the mean field potential to the contribution of the poling field, it reads

$$U_T \approx U_E = -\vec{\mu}_0^*.\vec{E}_P = -\left[\frac{\varepsilon(n^2+2)}{(n^2+2\varepsilon)}\vec{\mu}_0\right].\vec{E}_P \qquad (3.48)$$

where μ_0^* is the molecular dipole moment corrected from the local field factor given by the Onsager expression (static local field correction factor;

Böttcher 1973), μ is the molecular ground state dipole moment, ε is the static dielectric constant, and n the refractive index. In this description, the bulk response of oriented materials at room temperature is calculated by considering the molecular ensemble in the high temperature state, locked into the orientation obtained by the electric field. Taking Equation 3.48 into Equation 3.47 gives

$$G(\Omega) = \frac{\exp\left(\frac{\vec{\mu}_0^* . \vec{E}_P}{k_B T}\right)}{\int d\Omega \exp\left(\frac{\vec{\mu}_0^* . \vec{E}_P}{k_B T}\right)} \tag{3.49}$$

The efficiency of the poling procedure depends on the ratio of the electrostatic dipole alignment energy to thermal energy. This ratio is represented by the dimensionless parameter u^*:

$$u^* = \frac{\vec{\mu}_0^* . \vec{E}_P}{k_B T} \tag{3.50}$$

In conclusion, bulk susceptibilities are statistical averages of the molecular polarizabilities affected by the surroundings (i.e., corrected for local field effects). The orientations of the molecular moieties affects the relative magnitude of each bulk tensor component and are taken into account by evaluating the orientational average with respect to a given distribution function $G(\vec{\Omega})$. Under the assumption that the molecules are noninteracting, the bulk response is simply proportional to the concentration of molecular units, N. We will see in the next section an application of the dilute model in the case of SHG coefficients in dye-doped polymers, even at high chromophore loadings. Note that, if we apply an isotropic averaging through $G(\vec{\Omega})$, then we are left with an average isotropic expression that should give no even response in the case of achiral moieties ($P_I^{(0)} = 0, \chi_{IJK}^{(2)} = 0$).

Remark: Coherent response versus incoherent response. *SHG or SFG are coherent processes that collect a NLO response with a phase delay between the input field and output field when propagating in a medium. Thus, this means that the collected intensity gives direct access to the field amplitude and to the susceptibility tensor components that reflect the local and permanent symmetry of the analyzed medium (time average). Here, the intensity is proportional to the square of the concentration of molecular units N^2. However, when collecting scattered light, like hyper-Rayleigh, the problem is quite different since the process is incoherent. In that case, no explicit phase delay is measurable since the collected* *intensity originates from time and spatial fluctuations of the induced local polarization, that is, it relates to the autocorrelation function of the molecular polarization $<p(0)p(t)>$. In that sense, under the hypothesis of noninteracting moieties, an isotropic medium gives a response that reflects the molecular symmetry (see Chapter 2). Here, the intensity is proportional to the concentration of molecular units (N). Thus, a centrosymmetric molecule will give no hyper-Rayleigh scattering whereas a polar moiety will produce a hyper-Rayleigh scattering signal even if the medium is statistically isotropic.*

3.4.2 Application to an uniaxial polar interface

3.4.2.1 Experimental ODF-based approach

The hyperpolarizability tensor of push-pull chromophores is usually dominated by a single longitudinal coefficient $\beta_{zzz} = \beta$ associated with the push-pull molecular axis (see Chapter 2). Therefore, we can assume a cylindrical symmetric chromophore, that is, an average of the hyperpolarizability over the ψ Euler angle. To complete the uniaxial symmetry at the interface ($C_{\infty v}$ symmetry), we have to average the distribution over the ϕ Euler angle around the normal axis of the film. Thus, the ODF for a uniaxial interface containing cylindrically symmetric probe chromophores is completely defined by the polar angle (θ) between the unique long symmetry axis of the dye and the laboratory vertical axis (Figure 3.27).

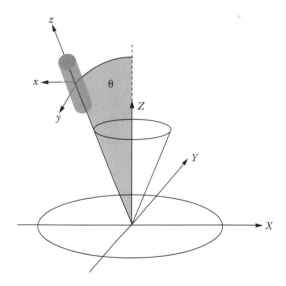

Figure 3.27 Schematic diagram of planar polar uniaxial interface obtained from a statistical ODF $G(\theta)$ of a cylindrical push-pull molecule.

From Chapter 1, there are four non-vanishing tensor components for an uniaxial system with $C_{\infty v}$ symmetry (Table 1.1):

$$
\begin{bmatrix}
0 & 0 & 0 & 0 & 0 & xzx & xxz & 0 & 0 \\
0 & 0 & 0 & xxz & xzx & 0 & 0 & 0 & 0 \\
zxx & zxx & zzz & 0 & 0 & 0 & 0 & 0 & 0
\end{bmatrix}
\tag{3.51}
$$

Thus, for the degenerate SHG case ($\chi^{(2)}_{XZX} = \chi^{(2)}_{XXZ}$), we are left with three nonzero susceptibility components that can be calculated following Equation 3.46 as

$$
\chi^{(2)}_{ZZZ} = N\beta^{*}_{zzz}\langle \cos^{3}\theta \rangle
$$

$$
\chi^{(2)}_{ZXX} = N\beta^{*}_{zzz}\frac{1}{2}\langle \cos\theta \sin^{2}\theta \rangle = N\beta^{*}_{zzz}\frac{1}{2}(\langle \cos^{3}\theta \rangle - \langle \cos\theta \rangle)
\tag{3.52}
$$

$$
\chi^{(2)}_{XXZ} = \chi^{(2)}_{ZXX}
$$

where the brackets $\langle\rangle$ indicate an average over the molecular orientation weighted by the ODF $G(\theta)$ as

$$
\langle \cos^{n}\theta \rangle_{I} = \frac{\int_{-1}^{+1} d(\cos\theta)G(\theta)\cos^{n}\theta}{\int_{-1}^{+1} d(\cos\theta)G(\theta)}
\tag{3.53}
$$

From Equation 3.52, we immediately see that assuming a single longitudinal coefficient $\beta_{zzz} = \beta$ for the chromophore implies the equality $\chi^{(2)}_{XXZ} = \chi^{(2)}_{ZXX}$ on the resulting bulk components for a uniaxial symmetry, which is in fact a pseudo Kleinman condition.

We now aim to express the orientational distribution function in terms of measurable quantities that describe the bulk organization. Whatever the mean field potential U_T depicted in the preceding section (Equation 3.47), the angular distribution function $G(\theta)$ may always be developed on the basis of the Legendre polynomials:

$$
G(\theta) = \sum_{i=0}^{\infty}\left(i+\frac{1}{2}\right)\langle P_{i} \rangle P_{i}(\cos\theta)
\tag{3.54}
$$

where only the even-order polynomials are contributing in centrosymmetric systems, since $G(\theta) = G(\pi - \theta)$, but both sets of odd- and even-order parameters can be effective in noncentrosymmetric interfaces. The order parameters or coefficients $\langle P_i \rangle$ are given by:

$$\langle P_i \rangle = \int_{-1}^{+1} d(\cos\theta) P_i(\cos\theta) G(\theta) \tag{3.55}$$

with

$$\langle P_0 \rangle = 1$$

$$\langle P_1 \rangle = \langle \cos\theta \rangle$$

$$\langle P_2 \rangle = \frac{1}{2}(3\langle \cos^2\theta \rangle - 1)$$

$$\langle P_3 \rangle = \frac{1}{2}(5\langle \cos^3\theta \rangle - 3\langle \cos\theta \rangle)$$

$$\langle P_4 \rangle = \frac{1}{8}(35\langle \cos^4\theta \rangle - 30\langle \cos^2\theta \rangle + 3) \tag{3.56}$$

The development in Equation 3.54 is generally restricted to the first known $\langle P_i \rangle$ values and, obviously, it is not mandatory that the series is converging.

Now taking Equation 3.56 into Equation 3.52 gives

$$\chi_{ZZZ}^{(2)} = N\beta_{zzz}^* \left(\frac{3}{5}\langle P_1 \rangle + \frac{2}{5}\langle P_3 \rangle \right)$$

$$\chi_{ZXX}^{(2)} = N\beta_{zzz}^* \left(\frac{1}{5}\langle P_1 \rangle - \frac{1}{5}\langle P_3 \rangle \right) = \chi_{ZXX}^{(2)} \tag{3.57}$$

The ratio $\chi_{ZZZ}^{(2)}/\chi_{ZXX}^{(2)}$ can be derived from Equation 3.57, independent of the molecular hyperpolarizability (β_{zzz}). It reads

$$\frac{\chi_{ZZZ}^{(2)}}{\chi_{ZXX}^{(2)}} = \frac{3\langle P_1 \rangle + 2\langle P_3 \rangle}{\langle P_1 \rangle - \langle P_3 \rangle} \tag{3.58}$$

Assuming a broad polar distribution where $\langle P_1 \rangle \gg \langle P_3 \rangle \approx 0$, we obtain $\chi_{ZZZ}^{(2)}/\chi_{ZXX}^{(2)} \approx 3$. Indeed, in thin-poled films, the value of this ratio roughly scales the SHG experimental ratio $\sqrt{I_{pp}^{2\omega}/I_{sp}^{2\omega}}$ (Brasselet and Zyss 1996).

Now, combining the two relations of Equation 3.57, we can express the first two odd-order parameters, $\langle P_1 \rangle$ and $\langle P_3 \rangle$, from the two bulk tensor components experimentally determined, scaled by the density number

(N) and the molecular hyperpolarizability (β_{zzz}^*):

$$\langle P_1 \rangle = \frac{1}{N\beta_{zzz}^*}\left(\chi_{ZZZ}^{(2)} + 2\chi_{ZXX}^{(2)}\right)$$

$$\langle P_3 \rangle = \frac{1}{N\beta_{zzz}^*}\left(\chi_{ZZZ}^{(2)} - 3\chi_{ZXX}^{(2)}\right)$$

(3.59)

In Section 3.3, from ultraviolet-visible spectra of doped polymer with push-pull chromophores, we have introduced the order parameter Φ defined by $\Phi = 1 - A_\perp/A_0$, where A_0 and $A_\perp$ are, respectively, the absorbance maxima for the unpoled and poled samples at normal incidence. In fact, under the uniaxial symmetry assumption, following Equation 3.56 it can be demonstrated that (Cao and McHale 1997):

$$\langle P_2 \rangle = \frac{A_{//} - A_\perp}{A_{//} + 2A_\perp} \approx \frac{A_0 - A_\perp}{A_0} = \Phi$$

(3.60)

with $A_{//}$ and $A_\perp$ in the first part of the formula the in- and out-of-plane absorption determined from angle dependent UV/V is absorption experiments.

It is worth noting that even-order parameters can be accessed through other optical techniques such as infrared absorption (IR: $\langle P_2 \rangle$) or Raman scattering (RS: $\langle P_2 \rangle, \langle P_4 \rangle$; Lagugné-Labarthel et al. 2004). More extensively, a few experimental techniques give additional clues to solve the ODF of surfaces (Figure 3.28). Nevertheless, SHG is the only optical technique that

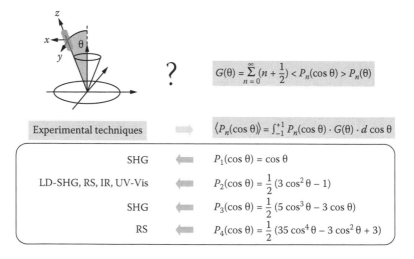

Figure 3.28 Selected experimental techniques that probe the first four order parameters of the uniaxial ODF $G(\theta)$ of a cylindrical push-pull molecule.

gives a rather direct access to the first two odd-order parameters, that is, it addresses an indication of the amplitude of the polar and/or octupolar order. When including anisotropic complex linear optical constants (index of refraction and absorption; Rodriguez 2008), in strong absorbing films it is also possible to gain linear dichroism information $(\langle P_2 \rangle)$ from SHG experiments (we will refer to this procedure as LD-SHG, not be confused with SHG-LD discussed in the next chapter) from the harmonic wave propagation, which is usually attenuated by the charge-transfer band absorption (Rodriguez et al. 2003).

Additionally, the maximum entropy criterion can be applied, which determines the most probable ODFs. It consists in maximizing the statistical entropy of the ODF, defined as follows:

$$S(G(\theta)) = -\int_{-1}^{+1} d(\cos\theta) G(\theta) Ln(G(\theta)) \qquad (3.61)$$

We will not detail the procedure, but further indications and references are given elsewhere (Rodriguez et al. 2005).

Then, an accurate and elaborated procedure to determine an uniaxial ODF from "push-pull" chromophores at an interface would be:

1. Quantify the one-dimensional (1-D) hyperpolarizability β_{zzz} (with hyper-Rayleigh scattering measurements, for example).
2. Quantify the non-vanishing susceptibility tensor components (Equation 3.59) obtained from accurate polarized SHG experiments.
3. Obtain ultraviolet-visible spectra before and after poling that gives axial information using Equation 3.60 (or any linear dichroism measurements like LD-SHG).
4. Build the truncated ODF $(G'(\theta))$ up to the third order from points 1 to 3 (Equation 3.54): $G'(\theta) = \frac{1}{2} + \frac{3}{2} < P_1 > + \frac{5}{2} < P_2 > + \frac{7}{2} < P_3 > \approx G(\theta)$.
5. Eventually, apply maximum entropy information either as a last step to obtain the most probable ODF $(G''(\theta) \approx G(\theta))$ or as an intermediate condition to Step 2, through an iterative converging procedure, to determine the most probable tensor components that maximize the orientational entropy.

As a worked out example of this procedure, let us consider a disperse red one (DR1) co-MMA polymer (12% molar). DR1 is a well-known dye with a 1-D charge-transfer character (ideal push-pull chromophore). A thin spin-coated film with a thickness of 500 nm has been poled following a classical procedure (3 kV during 30 minutes at 80°C). The number density was $N \approx 7\ 10^{26}$ *chromophores/m³* and the film exhibited a strong absorption band at 470 nm. As a consequence, the harmonic wave of a

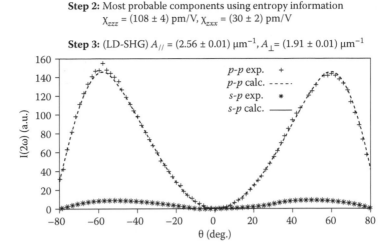

Step 2: Most probable components using entropy information
$\chi_{zzz} = (108 \pm 4)$ pm/V, $\chi_{zxx} = (30 \pm 2)$ pm/V

Step 3: (LD-SHG) $A_{//} = (2.56 \pm 0.01)\ \mu m^{-1}$, $A_{\perp} = (1.91 \pm 0.01)\ \mu m^{-1}$

Figure 3.29 SHG experimental (points) and calculated (solid lines) p–p and s–p polarized patterns. Calculated patterns were obtained from both the best fit to experimental curves and maximum entropy information.

1064 nm excitation line was strongly absorbed and LD-SHG could be performed.

As the first step, the longitudinal 1-D hyperpolarizability of DR1 at 1064 nm has been determined by Kaatz and Shelton (1996); $\beta_{zzz} = 2950 \times 10^{-40}\ m^4/V$.

Steps 2 and 3 are summarized in Figure 3.29, which reports standard SHG polarized experiments that have been treated using a general ellipsometric method, including maximum entropy criterion (Rodriguez et al. 2005), and that also includes linear optical anisotropic absorption (LD-SHG) as a simultaneous Step 3. Ultraviolet-visible experiments that were performed together gave the same information and confirmed LD-SHG results from Step 3.

Steps 4 and 5 are summarized together in Figure 3.30. The order parameters were obtained through an iterative procedure including maximum entropy information. The truncated ODF ($G'(\theta)$ up to the third order) could be obtained and the final best set of order parameters, including entropy information, is reported. The final step was to obtain the most probable ODF $G''(\theta)$, which is reported in Figure 3.30. The ODF is typical of a quite efficient poling, but still with $\langle P_3 \rangle$ close to zero. Besides, the axial anisotropy with $\langle P_2 \rangle \approx 0.10$ is still smaller than liquid crystal ordering, where typical values are 0.6–0.8, but it indicates that the intermolecular forces are strong enough to introduce such anisotropy. In fact, when the poling is efficient, we expect a strong increase of the anisotropy not only for the odd terms but also for the even-order parameters. This point

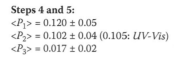

Steps 4 and 5:
$\langle P_1 \rangle = 0.120 \pm 0.05$
$\langle P_2 \rangle = 0.102 \pm 0.04 \ (0.105: \ UV\text{-}Vis)$
$\langle P_3 \rangle = 0.017 \pm 0.02$

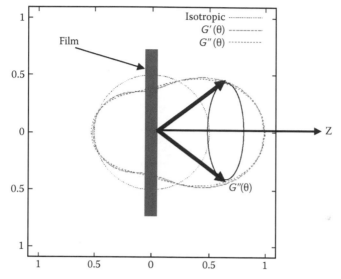

Figure 3.30 Order parameters obtained from SHG/ultraviolet-visible experiments and polar plot of the corresponding ODF, where G′(θ) is the truncated distribution obtained from the order parameters and G″(θ) is the most probable distribution.

has been well established in the case of the homopolymer p(DR1M) (100% molar; Rodriguez et al. 2003, 2005).

3.4.2.2   Mean-field potential approach: The liquid-oriented model

As demonstrated many times in the literature, in poled polymers the chromophore orientation during the poling process usually does not exhibit a strong orientation; it follows much more the oriented-liquid model that we detail now. We consider the ODF of an ensemble of freely rotating dipoles under the influence of a static electric field. We have seen that the corresponding Gibbs distribution is given by Equation 3.49 where the dimensionless parameter u^* (see Equation 3.50) represents the ratio of the dipole alignment work energy to thermal energy. Substituting Equation 3.49 into Equation 3.46 and evaluating the orientational average to first order in $u^* (u^* \ll 1)$ yields

$$\chi_{IJK}^{(2)} = N \left\langle \beta_{ijk}^* \right\rangle_{IJK} = N \int_0^{2\pi} d\phi \int_0^{2\pi} d\psi \int_{-1}^{+1} d(\cos\theta) \, \beta_{ijk}^* \, a_{Ii} \, a_{Jj} \, a_{Kk} \, u^* \tag{3.62}$$

Assuming again a one-dimensional molecule, that is the only non-vanishing component of μ_0^* and β^* is $\mu_{0,z}^*$ and β_{zzz}^*, respectively, the two independent bulk components are

$$\chi_{ZZZ}^{(2)} = N\beta_{zzz}^* \frac{\mu_{0,z}^* E_P}{5k_BT}$$

$$\chi_{ZXX}^{(2)} = N\beta_{zzz}^* \frac{\mu_{0,z}^* E_P}{15k_BT}$$

(3.63)

Here, we recognize Equation 3.36, which had been introduced in Section 3.3 as a $\chi^{(2)}$ orientational contribution. In fact, the oriented-liquid model or (Langevin model) has been largely used in this field to guide the development of poled materials. Then in the first-order approximation, a ratio $\chi_{ZZZ}^{(2)}/\chi_{ZXX}^{(2)} = 3$ can be derived directly from Equation 3.63, independent of the molecular quantities β_{zzz}^* and m_z^*.

Indeed, more extended models have been considered in the literature, such as poled material with axial order with or without strong poling fields. A complete and still up-to-date review of these considerations can be found in Kuzyk and Dirk (1998).

3.5 Surface adsorption and surface reactions

Catalysis is an essential instrument to make chemical technology more efficient and competitive. Progress in catalysis is largely based on a better understanding of molecular interactions between catalyst, reagents, and reaction products. For this purpose, spectroscopic and surface sensitive techniques with good time resolution are essential. While standard techniques such as atomic force microscopy (AFM), scanning tunneling microscopy (STM), and transmission electron microscopy (TEM) can provide spatial images of the catalyst, second-harmonic generation is able to probe changes in molecular structure and symmetry with both temporal and spatial resolution.

For the case of a layer of adsorbates on an interface or surface, the total nonlinear susceptibility of the interface can be expressed as:

$$\chi_{total}^{(2)} = \chi_{substrate}^{(2)} + \chi_{int}^{(2)} + \chi_{ad}^{(2)}$$

(3.64)

with $\chi_{substrate}^{(2)}$, $\chi_{int}^{(2)}$, and $\chi_{ad}^{(2)}$ denoting contributions from the bare substrate, the adsorbate–substrate interaction, and the adsorbate layer, respectively. Which of these terms dominates is dependent on the system under study. For example, in the case of adsorption of a molecule with high hyperpolarizability, $\chi_{ad}^{(2)}$ will probably dominate the nonlinear optical response. In general, however, all terms can contribute. Furthermore, since the susceptibilities are usually complex-valued, this can result in a very complicated behavior of the SHG response upon adsorption.

Whereas the term $\chi_{substrate}^{(2)}$ will be constant during the adsorption process, the other two terms are time-dependent since they are connected to the amount of adsorbate on the substrate. Hence, by monitoring the SHG signal versus time it will be possible to deduce information about the adsorption dynamics. However, to extract quantitative information about the adsorption kinetics, the explicit time-dependence of the susceptibilities needs to be known. Since this will depend on the particular system under study, it will be useful to consider a simple example.

If we consider the system

$$S + A \underset{k_d}{\overset{k_{ad}}{\rightleftharpoons}} S - A \tag{3.65}$$

where the adsorbate (A) adsorbs onto the substrate (S) to form a substrate–adsorbate complex ($S-A$). Both adsorption and desorption are considered with rate constant k_{ad} and k_d, respectively. The overall rate equation is given by

$$\frac{dC_s}{dt} = -k_{ad}C_s + k_d C_{S-A} \tag{3.66}$$

where C_i represents surface concentrations. If we now rewrite this equation in terms of surface coverage (Θ) and number of available adsorption sites (N_S), with

$$C_S = (1 - \Theta)N_s$$
$$C_{S-A} = \Theta N_s \tag{3.67}$$

we obtain

$$\frac{d\Theta}{dt} = -(k_{ad} + k_d)\Theta + k_{ad} \tag{3.68}$$

The solution of this differential equation is

$$\Theta(t) = \frac{k_{ad} - k_{ad}e^{-(k_{ad}+k_d)t}}{k_{ad} + k_d} \tag{3.69}$$

If we now assume that the overall surface susceptibility is dominated by $\chi_{ad}^{(2)}$, we can write the second-harmonic intensity as

$$I(2\omega) = K\left|\chi_{ad}^{(2)}\right|^2 I^2(\omega) = K\left|N_s\Theta(t)f\beta\right|^2 I^2(\omega) \tag{3.70}$$

with β the hyperpolarizability of the adsorbate and f a local field correction factor.

Hence,

$$\Theta(t) = \frac{\sqrt{I(2\omega)}}{KN_s\left|\beta\right| fI(\omega)} = K'\sqrt{I(2\omega)} \tag{3.71}$$

or

$$\sqrt{I(2\omega)} = K''[1 - e^{-(k_{ad}+k_d)t}] \tag{3.72}$$

and surface coverage versus time can be monitored by recording the square root of the SH intensity versus time. A simulation is shown in Figure 3.31.

The situation becomes more complicated when $\chi_{substrate}^{(2)}$ and $\chi_{ad}^{(2)}$ contribute. Since both susceptibilities may have different signs and phases,

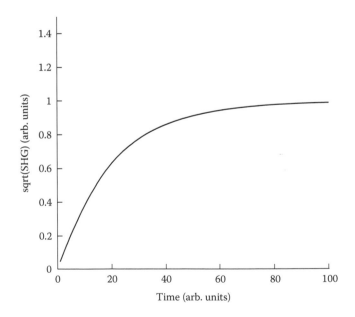

Figure 3.31 $\sqrt{I(2\omega)}$ versus time for the particular situation where the second-harmonic response is dominated by $\chi_{ad}^{(2)}$.

adsorption does not necessarily imply a rise in the SHG signal. If we take into account real (re) and imaginary (im) parts of the surface susceptibilities, Equation 3.70 transforms into

$$I(2\omega) = K \left| \chi_{substrate}^{(2)} + \chi_{ad}^{(2)} \right|^2 I^2(\omega) = K \left| \chi_{substrate}^{(2)} + N_s \Theta(t) f\beta \right|^2 I^2(\omega)$$

$$= K \left| \chi_{re,substrate}^{(2)} + i\chi_{im,substrate}^{(2)} + N_s \Theta(t) f(\beta_{re} + i\beta_{im}) \right|^2 I^2(\omega) \qquad (3.73)$$

$$= K \left| (\chi_{re,substrate}^{(2)} + N_s \Theta(t) f\beta_{re}) + i(\chi_{im,substrate}^{(2)} + N_s \Theta(t) f\beta_{im}) \right|^2 I^2(\omega)$$

As we can see, the situation becomes quite complex. In the most general case one would also have to include the interaction term. In Figure 3.32 we simulate a few situations that could occur.

It is important to note that the previous discussion is only valid for the simple system given in Equation 3.65. In reality, adsorption processes can be quite complex and surface reactions may occur. In that case, experimental results may be difficult to interpret.

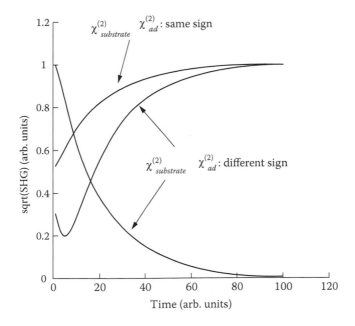

Figure 3.32 $\sqrt{I(2\omega)}$ versus time for the particular situation where the second-harmonic response is due to contributions from $\chi_{ad}^{(2)}$ and $\chi_{substrate}^{(2)}$.

References

Abéles, F., *Optical Properties of Solids*, Amsterdam: North-Holland, 1972.

Alley, T.G., S.R.J. Brueck, and M. Wiedenbeck, *J. Appl. Phys.*, 86 (1999), 6634.

Azzam, R.M.A., and N.M. Bashara, *Ellipsometry and Polarized Light*, Amsterdam: North-Holland, 1977.

Berreman, D.W., *J. Opt. Soc. Am.*, 62 (1972), 502.

Bethune, D.S., *J. Opt. Soc. Am. B*, 6 (1989), 910.

Bethune, D.S., *J. Opt. Soc. Am. B*, 8 (1991), 367.

Bloembergen, N., and P.S. Pershan, *Phys. Rev.*, 128 (1962), 606.

Bosshard, C., U. Gubler, P. Kaatz, W. Mazerant, and U. Meier, *Phys. Rev. B*, 61 (2000), 10688.

Böttcher, C.J.F., *Theory of Electric Polarization*, 2nd ed., Amsterdam: Elsevier, 1973.

Brasselet, S., and J. Zyss, *J. of Nonlinear Opt. Phys. and Mat.*, 5 (1996), 671.

Braun, M., F. Bauer, T. Vogtmann, and M. Schwoerer, *J. Opt. Soc. Am. B*, 14 (1997), 1699.

Braun, M., F. Bauer, T. Vogtmann, and M. Schwoerer, *J. Opt. Soc. Am. B*, 15 (1998), 2877.

Burland, D.M., R.D. Miller, and C.A. Walsh, *Chem. Rev.*, 94 (1994), 31.

Butcher, P.N., and D. Cotter, *The Elements of Nonlinear Optics* (Cambridge Studies in Modern Optics, vol. 9), Cambridge, U.K.: Cambridge University Press, 1990.

Cao, X., and J.L. McHale, *J. Phys. Chem. B*, 101 (1997), 8843.

Cardinal, T., E. Fargin, G. Le Flem, and S. Leboiteux, *J. of Non-Cryst. Sol.*, 222 (1997), 228.

Cardinal, T., E. Fargin, G. Le Flem, et al., *Eur. J. Solid State Inorg. Chem.*, t.33 (1996), 597.

Chemla, D., and P. Kupecek, *Phys. Rev. Appl.*, 6 (1971), 31.

Chemla, D.S., and J. Zyss, *Nonlinear Optical Properties of Organic Molecules and Crystals* (Quantum Electronics: Principles and Applications Series, vol. 1), London: Academic Press, 1987.

Dechter, G., B. Tieke, C. Bosshard, and P. Gunter, *J. Chem. Soc. Chem. Commun.*, (1988), 933.

Dmitriev, V.G., G.G. Gurzadyan, and D.N. Nikogosyan, *Handbook of Nonlinear Optical Crystal*, 2nd ed., Berlin: Springer, 1997.

Dussauze, M., E. Fargin, M. Lahaye, V. Rodriguez, and F. Adamietz, *Opt. Express*, 13 (2005), 4064.

Eisenthal, K.B., *Chem. Rev.*, 96 (1996), 1343.

Enami, Y., C.T. DeRose, D. Mathine, C. Loychik, C. Greenlee, R.A. Norwood, T.D. Kim, et al., *Nature Photonics*, 1 (2007), 180.

Guyot-Sionnest, P., W. Chen, and Y.R. Shen, *Phys. Rev.*, B33 (1986), 8254.

Henry, L.J., B.V. McGrath, T.G. Alley, and J.J. Kester, *J. Opt. Soc. Am. B*, 13 (1996), 827.

Herman, W., and L.M. Hayden, *J. Opt. Soc. Am. B*, 12 (1995), 416.

Imai, H., S. Horinouchi, N. Asakuma, K. Fukao, D. Matsuki, H. Hirashima, and K. Sasaki, *J. Appl. Phys.*, 84 (1998), 5415.

Jerphagnon, J., and S.K. Kurtz, *J. Appl. Phys.*, 41 (1970), 1667.

Kaatz, P., and D.P. Shelton, *J. Chem. Phys.*, 105 (1996), 3918.

Kao, Y.H., Y. Hu, H. Zheng, J.D. Mackenzie, K. Perry, G. Bourhill, and J.W. Perry, *J. Non-Cryst. Solids*, 167 (1994), 247.

Kazansky, P.G., A. Kamal, and P. St. J. Russel, *Opt. Lett.*, 18 (1993), 693.

Kazansky, P.G., and P. St. J. Russell, *Opt. Commun.*, 101 (1994), 611.

Kim, T.-D., J.-W. Kang, J. Luo, S.-H. Jang, J.-W. Ka, N. Tucker, J.B. Benedict, et al., *J. Am. Chem. Soc.*, 129 (2007), 488.

Kudlinski, A., Y. Quiquempois, and G. Martinelli, *Opt. Lett.*, 30 (2005a), 1039.

Kudlinski, A., Y. Quiquempois, and G. Martinelli, *Appl. Phys. Lett.*, 86 (2005b), 161909.

Kuzyk, M.G., and C.W. Dirk, *Characterization Techniques and Tabulations for Organic Nonlinear Optical Materials*, New York: Marcel Dekker, 1998.

Kuzyk, M.G., K.D. Singer, H.E. Zahn, and L.A. King, *J. Opt. Soc. Am. B*, 6 (1989), 742.

Lagugné-Labarthet, F., J.L. Bruneel, V. Rodriguez, and C. Sourisseau, *J. Phys. Chem. B*, 108 (2004), 1267.

Lee, S.-S., S.M. Garner, V. Chuyanov, H. Zhang, W.H. Steier, F. Wang, L.R. Dalton, A.H. Udupa, and H.R. Fetterman, *IEEE J. Quant. Elect.*, 36 (2000), 527.

Lindsay, G.A., and K.D. Singer, Polymer for Second-Order Nonlinear Optics, ACS Symposium Series 601, Washington, D.C., 1995.

Maker, P.D., R.W. Terhune, M. Nisenoff, and C.M. Savage, *Phys. Rev. Lett.*, 8 (1962), 21.

Mortazavi, M.A., A. Knoesen, S.T. Kowel, B.G. Higgins, and A. Dienes, *J. Opt. Soc. Am. B*, 6 (1989), 733.

Mukherjee, N., R.A. Myers, and S.R.J. Brueck, *J. Opt. Soc. Am. B.*, 11 (1994), 665.

Myers, R.A., N. Mukherjee, and S.R.J. Brueck, *Opt. Lett.*, 16 (1991), 1732.

Provencher, P., M.-M. Denarieze Roberge, A. Suau, K. Tian, G. Munger, and R. Leblanc, *J. Opt. Soc. Am.*, 12 (1995), 1406.

Qui, M., S. Egawa, K. Horimoto, and T. Mizunami, *Opt. Commun.*, 189 (2001), 161.

Quiquempois, Y., G. Martinelli, P. Dutherage, P. Bernage, P. Niay, and M. Douay, *Opt. Commun.*, 176 (2000), 479.

Quiquempois, Y., N. Godbout, and S. Lacroix, *Phys. Rev. A*, 65 (2002), 043816.

Robinson, B.H., L.R. Dalton, A.W. Harper, A. Ren, F. Wang, C. Zhang, G. Todorova, et al., *Chem. Phys.*, 245 (1999), 35.

Rodriguez, V., *J. Chem. Phys.*, 128 (2008), 064707.

Rodriguez, V., F. Adamietz, L. Sanguinet, T. Buffeteau, and C. Sourisseau, *J. Phys. Chem. B*, 107 (2003), 9736.

Rodriguez, V., F. Lagugné-Labarthet, and C. Sourisseau, *Appl. Spectrosc.*, 59 (2005), 322.

Rodriguez, V., and C. Sourisseau, *J. Opt. Soc. Am.*, B19 (2002), 2650.

Sasaki, Y., and Y. Ohmori, *Appl. Phys. Lett.*, 39 (1981), 466.

Schrader, S., V. Zauls, B. Dietzel, C. Flueraru, D. Prescher, J. Reiche, H. Motschmann, and L. Brehmer, *Mater. Sci. Eng. C*, 8-9 (1999), 527.

Schwartz, H., R. Mazor, V. Khodorkovsky, L. Shapiro, J.T. Klug, E. Kovalev, G. Meshulam, G. Berkovic, Z. Kotler, and S. Efrima, *J. Phys. Chem. B*, 105 (2001), 5914.

Shen, Y.R., *The Principles of Nonlinear Optics*, New York: Wiley & Sons, 1984.

Shoji, I., T. Kondo, A. Kitamoto, M. Shirane, and R. Ito, *J. Opt. Soc. Am. B*, 14 (1997), 2268.

Sigaev, V.N., S. Yu. Stefanovich, B. Champagnon, I. Gregora, and P. Pernice, *J. Non-Cryst. Solids*, 306 (2002), 238.

Singer, K.D., M.G. Kuzyk, and J.E. Sohn, *J. Opt. Soc. Am. B*, 4 (1987), 968.

Singer, K.D., J.E. Sohn, and S.J. Lalama, *Appl. Phys. Lett.*, 49 (1986), 248.

Von Hippel, A., E.P. Gross, J.G. Jelatis, and M. Geller, *Phys. Rev.*, 91 (1953), 568.

Wijekoon, W.M.K.P., S.P. Karna, G.B. Talapatra, and P.N. Prasad, *J. Opt. Soc. Am.*, 10 (1993), 213.

Wijekoon, W.M.K.P., S.K. Wijaya, J.D. Bhawalkar, P.N. Prasad, T.L. Penner, N.J. Armstrong, M.C. Ezenyilimba, and D.J. Williams, *J. Am. Soc.*, 118 (1996), 4480.

Yitzchaik, S., G. Berkovic, and V. Krongauz, *J. Appl. Phys.*, 70 (1991), 3949.

Yokoyama, S., T. Nakahama, A. Otomo, and S. Mashiko, *Thin Solid Films*, 331 (1998), 248.

chapter four

Characterization of surface chirality by second-harmonic generation and sum-frequency generation

4.1 Chirality and second-order nonlinear optics: general principles

4.1.1 Chirality and optical activity (Lakhtakia 1990)

Chiral molecules exist in two forms that are mirror images of each other, called enantiomers. Pasteur first demonstrated the existence of enantiomers in 1850 when he found the relation between chirality (or molecular dissymmetry as he called it) and optical activity studying sodium ammonium tartrate. Chirality is equivalent to the absence of mirror rotation axes of any order and chiral molecules exhibit the special characteristic that they can rotate the plane of polarization of linearly polarized light. This phenomenon had been discovered in 1812 by Biot and is known as optical rotation. At the origin of optical rotation lies a different refractive index for left- and right-hand circularly polarized light. A second tool to study chiral molecules is known as circular dichroism (discovered in 1895 by Cotton), which is the unequal absorption of left- and right-hand circularly polarized light. Both techniques are intensely used to study chiral molecules and systems.

Chiral molecules are of great scientific importance in biology and life sciences. Natural sugars and amino acids only exist in one enantiomeric conformation, although both enantiomers have the same energy and thus their formation has equal probability in an achiral environment. This problem has been intensely studied but nevertheless remains a mystery. The most important chiral structure is probably DNA, which receives much attention not in the least because of its impact on genetic engineering.

Also in the pharmaceutical industry, the importance of chiral molecules increases every year. In the past, many synthetic drugs were racemic mixtures, whereas those derived from natural products (for example, the penicillines) are almost always enantiomerically pure. Nowadays, the amount of enantiomerically pure drugs entering development is close to 100%. The reason is twofold: first, drugs become more complex and, second, a greater selectivity of action is sought. Often only one enantiomer of the drug has the wanted therapeutic effects, while the other is inactive, or, in the worst case, causes serious side effects.

To describe optical activity in terms of fundamental materials parameters such as polarizability or susceptibility, one has to also include magnetic-dipole contributions to the linear optical properties of the molecule (medium) by defining an induced dipole moment $\mu(\omega)$ (or polarization $P(\omega)$) that is also dependent on the magnetic induction field $B(\omega)$. Furthermore, we need to define an induced magnetic moment $m(\omega)$ (or magnetization $M(\omega)$). Both dipole moment (polarization) and magnetic moment (magnetization) act as sources of radiation. Then, up to first order in the magnetic-dipole interaction, we can write (Pattanayak and Birman 1981):

$$\mu(\omega) = \alpha^{ee} E(\omega) + \alpha^{em} B(\omega) \tag{4.1}$$

$$m(\omega) = \alpha^{me} E(\omega) \tag{4.2}$$

and

$$P(\omega) = \chi^{(1),ee} E(\omega) + \chi^{(1),em} B(\omega) \tag{4.3}$$

$$M(\omega) = \chi^{(1),me} E(\omega) \tag{4.4}$$

where α^{ee} is the usual linear polarizability and α^{em} and α^{me} are magnetic-dipole polarizabilities, with $\alpha^{em} = -\alpha^{me}$. The subscripts *ee*, *me*, and *em* are used to discriminate between the different polarizabilities and refer to electric-dipole (*e*) and magnetic-dipole (*m*) interactions. For example, the subscript *ee* indicates that a photon at frequency ω is annihilated through an electric-dipole interaction and a new photon is created through an electric-dipole interaction. For α^{em}, a photon at frequency ω is annihilated through a magnetic-dipole interaction and a new photon is created through an electric-dipole interaction. For α^{me}, a photon at frequency ω is annihilated through an electric-dipole interaction and a new photon is created through a magnetic-dipole interaction. $\chi^{(1),ee}$ is the usual electric-dipole susceptibility and $\chi^{(1),em}$ and $\chi^{(1),me}$ are magnetic-dipole susceptibilities, with $\chi^{(1),me} = -\chi^{(1),em}$. Note that for nonresonant excitation, $\chi^{(1),ee}$ is a real quantity, while $\chi^{(1),em}$ and $\chi^{(1),me}$ are imaginary quantities. It is important that α^{em}, α^{me}, $\chi^{(1),me}$, and $\chi^{(1),em}$ change signs between the enantiomers and

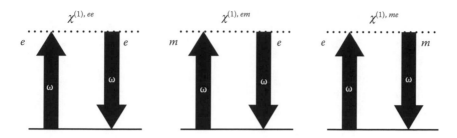

Figure 4.1 Energy diagrams for the susceptibilities $\chi^{(1),ee}$, $\chi^{(1),em}$, and $\chi^{(1),me}$.

vanish in achiral or racemic materials. The annihilation and creation of photons through the different mechanisms is illustrated in the energy diagrams of Figure 4.1.

The previous considerations allow us to intuitively understand optical rotation. For vertically polarized light incident on a chiral isotropic medium, a vertically directed polarization and magnetization will be created inside the medium. Both polarization and magnetization act as new sources of radiation: the polarization gives rise to a vertically polarized field, while the magnetization radiates a horizontally polarized field. The corresponding total field has its polarization rotated over an angle ϕ. The magnitude of ϕ depends on the relative magnitude of P and M. Furthermore, the rotation angle has the same magnitude but differs in sign for both enantiomers. In the absence of magnetic interaction, no optical rotation can occur. This is illustrated in Figure 4.2.

Circular dichroism, on the other hand, can be understood by considering circularly polarized light. Circularly polarized light, propagating

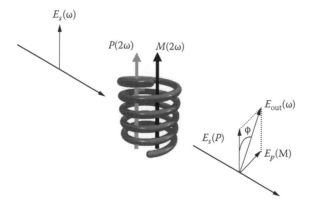

Figure 4.2 Intuitive view of optical activity.

along the z-direction, can be defined as

$$E^\pm = E_0(x \pm iy)(e^{ikz-i\omega t} + cc) \quad \text{and} \quad B^\pm = \mp iE^\pm \tag{4.5}$$

where the $\pm$ refers to left- and right-hand circular polarization, and x and y are unit vectors. If we now define an effective polarization of the form (see also Chapter 1; Shen 1984)

$$P_{eff} = P + i\frac{c}{\omega}\nabla \times M \tag{4.6}$$

and if we substitute Equations 4.3 and 4.4 into Equation 4.6 and use the Maxwell equations, it is easily seen that the effective polarization for left- and right-hand circularly polarized excitation ($P^\pm$) is equal to (Fisher and Hache 2005):

$$P_{eff}^\pm = \chi^{ee}E^\pm \mp 2i\chi^{em}E^\pm = \chi_{eff}^\pm E^\pm \tag{4.7}$$

Using the general equation for the refractive index, we obtain

$$n^\pm = \sqrt{1 + 4\pi\chi_{eff}^\pm} = \sqrt{1 + 4\pi\chi^{ee} \mp 8i\pi\chi^{em}} = \sqrt{n_0^2 \mp 8i\pi\chi^{em}} \tag{4.8}$$

with $n^\pm$ the refractive index for left- and right-hand circularly polarized light. Using a Taylor expansion further simplifies this expression into

$$n^\pm = n_0 \mp \frac{4i\pi\chi^{em}}{n_0} \tag{4.9}$$

As a result, the presence of the magnetic susceptibility explains the difference in refractive index for left- and right-hand circularly polarized light, which makes optical rotation possible.

For optical rotation, we need to consider the real part (Re) of the refractive indices n^+ and n^-, and then the rotation angle ϕ is given by:

$$\phi = \frac{\pi z}{\lambda}\text{Re}(n^+ - n^-) = \frac{-8\pi^2 z}{n_0\lambda}\text{Im}(\chi^{em}) \tag{4.10}$$

with z the distance traveled through the medium and $\text{Im}(\chi^{em})$ the imaginary part of χ^{em}. Note that it is the imaginary part of χ^{em} that is responsible for optical rotation. This is due to the fact that χ^{em} is a pure imaginary quantity away from resonance.

To explain circular dichroism one has to consider the imaginary part of the refractive index. If the optically active medium is not transparent at the wavelength of the incident radiation, then the transmitted intensity is substantially reduced, but not precisely the same amount as the isotropic absorption measured under the same conditions using unpolarized light. Not only do the two circularly polarized beams leave the material out of phase, but also one of the components is absorbed more strongly than the other. The resulting elliptically polarized light may be characterized by an angle ψ (the ellipticity ψ):

$$\psi = \frac{\pi z}{\lambda}(\mathrm{Im}(n^+ - n^-)) = \frac{8\pi^2 z}{n_0 \lambda} \mathrm{Re}(\chi^{em}) \tag{4.11}$$

In the more commonly used units for absorption, the extinction coefficient ε (*in* 1 *mol*$^{-1}$ *cm*$^{-1}$) is related to the imaginary part of the refractive index by:

$$\varepsilon = \frac{4\pi \, \mathrm{Im}\, n}{2,3\lambda C} \tag{4.12}$$

where C is the concentration in moles per liter. Substituting in Equation 4.11, the ellipticity in radians is

$$\psi = \frac{2,3Cz}{4}(\varepsilon^+ - \varepsilon^-) = \frac{32\pi^2}{2.3\lambda C n_0} \mathrm{Re}\, \chi^{em} \tag{4.13}$$

Instead of the ellipticity, one often uses the normalized quantity $\Delta\varepsilon/\varepsilon$ to quantify the circular dichroism, which is defined as:

$$\frac{\Delta\varepsilon}{\varepsilon} = \frac{2(\varepsilon^+ - \varepsilon^-)}{\varepsilon^+ + \varepsilon^-} = \frac{8\pi \, \mathrm{Re}\, \chi^{em}}{n_0 \, \mathrm{Im}\, n_0} \tag{4.14}$$

4.1.2   *Nonlinear optical activity in second-harmonic generation (Sioncke et al. 2003)*

Linear optical activity includes effects such as optical rotation and circular dichroism, as mentioned in the previous paragraph. In addition, materials also exhibit nonlinear optical activity, that is, a different response of the nonlinear optical (NLO) process to left- or right-hand circularly polarized light. Such circular difference effects have been observed in second-harmonic generation (SHG) generated in interfaces and films (Petralli-Mallow et al. 1993). For example, very large circular-difference effects have been observed in second-harmonic generation from chiral films. The effect can be seen as similar to ordinary circular dichroism (CD) and is therefore often referred to as SHG-CD. A nonlinear analogue of optical rotation has also been demonstrated experimentally and described theoretically (Byers et al. 1994). This effect measures the rotation in the polarization azimuth of the second-harmonic light generated at the surface with respect to that of the fundamental input light. Both SHG-CD and SHG-ORD (second-harmonic generation optical rotary dispersion) can be used to discriminate between enantiomers because the sign of the effect reverses for the other enantiomer. Furthermore, since SHG is surface sensitive and the nonlinear optical activity effects are specific for chiral materials, the effects can be used as a very sensitive tool to study chiral surfaces. Therefore, techniques based on nonlinear optical activity have been suggested for the study of biological membranes and interfaces.

As we have seen in previous chapters, SHG from interfaces and thin films can be described by the components of the nonlinear polarization $P_i(2\omega)$. In the electric-dipole approximation we have:

$$P_i(2\omega) = \sum_{j,k} \chi_{ijk}^{(2)} E_j(\omega) E_k(\omega) \tag{4.15}$$

Since the number of non-vanishing susceptibility components is directly linked to the symmetry of the sample and chirality is a symmetry property, we can expect a different susceptibility component when compared to achiral or racemic samples. Since most surfaces are isotropic in the plane of the sample, let us consider two symmetries, C_∞ and $C_{\infty v}$, where the former is a chiral sample with in-plane isotropy (a chiral isotropic surface) and the latter an achiral or racemic sample with in-plane isotropy (achiral isotropic surface). In Table 4.1 we list the number of non-vanishing components in both types of samples for the coordinate system shown in Figure 4.3.

As is evident from Table 4.1, the clear difference in terms of susceptibility components between both surfaces is the presence of an *xyz*-type susceptibility component in the chiral surface. Hence, measurement of this component would indicate the presence of chirality. Therefore, we will refer to this component as the chiral component, while the other susceptibility components are termed achiral components because they are present in chiral and achiral samples.

Table 4.1 Non-Vanishing Components of the Second-Order
Susceptibility Tensor for Second-Harmonic Generation for an Achiral
and Chiral Isotropic Surface

	Chiral Isotropic Surface	Achiral Isotropic Surface
Tensor components $\chi^{(2)}_{ijk}$	zzz	zzz
	$zxx = zyy$	$zxx = zyy$
	$xxz = yyz = yzy = xzx$	$xxz = yyz = yzy = xzx$
	$xyz = -yxz = -yzx = xzy$	

Now let us focus a little more on the symmetry properties of the chiral component. Since both enantiomers in chiral samples are connected by reflection symmetry, it seems justified to investigate how this component transforms under this symmetry operation (Figure 4.4). For a mirror plane in the zy-plane, the Cartesian coordinates transform as follows:

$$x \rightarrow -x \quad y \rightarrow y \quad z \rightarrow z$$

The effect of this symmetry operation is shown in Table 4.2

It is clear that only the sign of the chiral component changes upon reflection, which directly indicates that the chiral component has a different

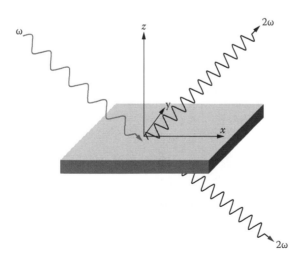

Figure 4.3 Coordinate system used to define the different non-vanishing susceptibility components.

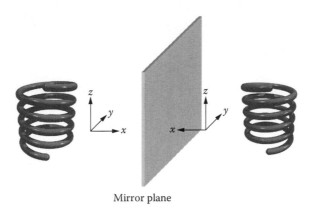

Mirror plane

Figure 4.4 Effect of a mirror plane on the Cartesian coordinates.

sign for both enantiomers. The other, achiral components remain unchanged. As a consequence, the sign of the chiral component can in principle be used to identify one enantiomer. This forms the basis of several SHG techniques that are able to probe chirality, which we will discuss next.

Without going into too much detail, it is relatively easy to intuitively understand optical rotation in second-harmonic generation from a chiral surface or thin film by simply considering the non-vanishing polarization components generated in a chiral and achiral sample film. For example, for an achiral sample thin film with $C_{\infty v}$ symmetry and for the experimental

Table 4.2 Effect of a Mirror Plane on the Sign of the Different Susceptibility Components

Before Symmetry Operation	After Symmetry Operation
Chiral Surface	
zzz	zzz
$zxx = zyy$	$zxx = zyy$
$xxz = yyz = yzy = xzx$	$xxz = yyz = yzy = xzx$
$\underline{xyz = -yxz = -yzx = xzy}$	$\underline{-xyz = yxz = yzx = -xzy}$
Achiral Surface	
zzz	zzz
$zxx = zyy$	$zxx = zyy$
$xxz = yyz = yzy = xzx$	$xxz = yyz = yzy = xzx$

Note: The chiral components are underlined to accentuate the changes in sign.

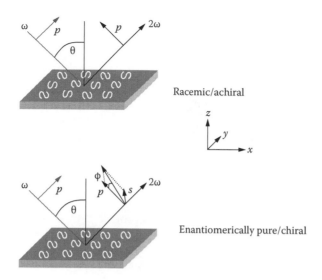

Figure 4.5 Schematic illustration of optical rotation in SHG.

situation shown in Figure 4.5 (fundamental beam *p*-polarized, SHG in reflection), the non-vanishing components of the polarization in the *xyz* coordinate system, can be written as:

$$P_z = \chi^{(2)}_{zzz}E_zE_z + \chi^{(2)}_{zxx}E_xE_x \qquad (4.16)$$

$$P_x = \chi^{(2)}_{xxz}E_xE_z + \chi^{(2)}_{xzx}E_zE_x \qquad (4.17)$$

Expressed in the *spk* coordinate system we obtain:

$$P_p(2\omega) = \left(\chi^{(2)}_{zzz}\sin^3\theta - 2\chi^{(2)}_{xxz}\cos^2\theta\sin\theta + \chi^{(2)}_{zxx}\cos^2\theta\sin\theta\right)E_p(\omega)E_p(\omega)$$

$$= f_p E_p(\omega)E_p(\omega) \qquad (4.18)$$

To simplify this equation we introduce the coefficient f_p, which is in this case is a combination of different susceptibility components in the *xyz* coordinate system, and the subscript *p*, which refers to the *p*-component of the nonlinear polarization. Later in this chapter we will introduce this coefficient in a more general fashion as a convenient way to describe non-linear optical processes.

From Equation 4.18 it is clear that a *p*-polarized fundamental beam will give rise to *p*-polarized second-harmonic generation: the polarization of the second-harmonic field is the same as that of the fundamental field for an achiral sample and no nonlinear optical rotation occurs.

When the sample becomes chiral (C_∞ symmetry), the components of the polarization are:

$$P_z = \chi^{(2)}_{zzz} E_z E_z + \chi^{(2)}_{zxx} E_x E_x \tag{4.19}$$

$$P_y = \chi^{(2)}_{yzx} E_z E_x + \chi^{(2)}_{yxz} E_x E_z \tag{4.20}$$

$$P_x = \chi^{(2)}_{xxz} E_x E_z + \chi^{(2)}_{xzx} E_z E_x \tag{4.21}$$

In the *spk* coordinate system this becomes:

$$P_p(2\omega) = \left(\chi^{(2)}_{zzz} \sin^3 \theta - 2\chi^{(2)}_{xxz} \cos^2 \theta \sin \theta + \chi^{(2)}_{zxx} \cos^2 \theta \sin \theta \right) E_p(\omega)E_p(\omega)$$
$$= f_p E_p(\omega)E_p(\omega) \tag{4.22}$$

$$P_s(2\omega) = \left(2\chi^{(2)}_{yzx} \cos \theta \sin \theta \right) E_p(\omega)E_p(\omega) = f_s E_p(\omega)E_p(\omega) \tag{4.23}$$

Again the coefficients f_s and f_p are combinations of susceptibility components defined in the *xyz* coordinate system, with the subscripts *s* and *p* referring to the *s*- and *p*-polarized component of the polarization second-harmonic light. It is easily seen that the presence of chirality introduces an additional polarization component in the s-direction. As a consequence, the polarization of the second-harmonic field will be rotated with respect to the polarization of the incident radiation. The amount of optical rotation will depend on the relative magnitude of the chiral and achiral susceptibility components. If we neglect linear optical parameters, the optical rotation is in this case given by

$$\phi = \tan^{-1} \frac{f_s}{f_p} = \tan^{-1} \frac{2\chi^{(2)}_{yzx} \cos \theta}{\chi^{(2)}_{zzz} \sin^2 \theta + 2\chi^{(2)}_{xxz} \cos^2 \theta + \chi^{(2)}_{zxx} \cos^2 \theta} \tag{4.24}$$

Furthermore, f_s changes signs between the enantiomers and consequently the optical rotation will change signs. Hence, this effect can be taken as a signature of chirality and is in a loose sense the nonlinear analogue of optical rotation. An important difference with linear optical rotation is that here we do not rely on magnetic contributions to explain optical rotation.

SHG-CD (Figure 4.6) can also be explained by considering all components of the polarization. For example, in the most general case (and for

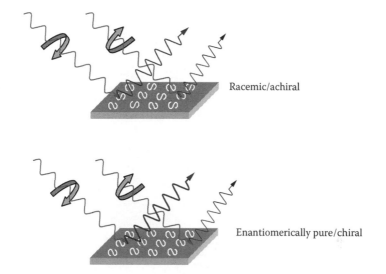

Figure 4.6 Schematic illustration of SHG-CD.

a reflection geometry), all relevant polarizability components in the *spk* coordinate system are

$$P_p(2\omega) = \left(\chi^{(2)}_{zzz} \sin^3 \theta - 2\chi^{(2)}_{xxz} \cos^2 \theta \sin \theta + \chi^{(2)}_{zxx} \cos^2 \theta \sin \theta \right) E_p(\omega) E_p(\omega)$$

$$+ \left(\chi^{(2)}_{zxx} \sin \theta \right) E_s(\omega) E_s(\omega) + \left(2\chi^{(2)}_{xyz} \cos \theta \sin \theta \right) E_s(\omega) E_p(\omega) \qquad (4.25)$$

$$= f_p E_p(\omega) E_p(\omega) + g_p E_s(\omega) E_s(\omega) + h_p E_s(\omega) E_p(\omega)$$

$$P_s(2\omega) = \left(2\chi^{(2)}_{xyz} \cos \theta \sin \theta \right) E_p(\omega) E_p(\omega) + \left(2\chi^{(2)}_{xxz} \cos \theta \sin \theta \right) E_s(\omega) E_p(\omega)$$

$$= f_s E_p(\omega) E_p(\omega) + h_s E_s(\omega) E_p(\omega) \qquad (4.26)$$

with coefficients f_i, g_i, and h_i again combinations of susceptibility components in the *spk* system. For circularly polarized input light, $E_p = \pm iE_s$ where the $\pm$ sign refers to left- and right-hand circular polarization. If we substitute this into Equations 4.25 and 4.26 we obtain

$$P_p(2\omega) = (-f_p + g_p \pm ih_p) E_s(\omega) E_s(\omega) \qquad (4.27)$$

$$P_s(2\omega) = (-f_s \pm ih_s) E_s(\omega) E_s(\omega) \qquad (4.28)$$

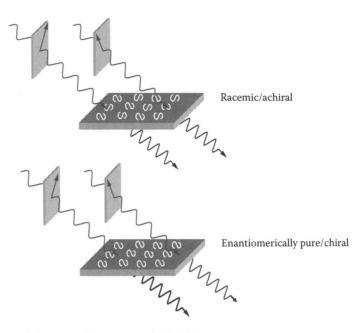

Figure 4.7 Schematic illustration of SHG-LD.

Hence, it is clear that both the *p*- and *s*-components of the nonlinear polarization can be different for left- and right-hand circular polarized excitation, and as a consequence a different efficiency in second-harmonic generation.

It is also quite interesting to note that +45° and −45° linearly polarized light can give rise to difference effects in SHG (Figure 4.7; Verbiest et al. 1995). For +45° and −45° linearly polarized light, $E_p = \pm E_s$, which yields

$$P_p(2\omega) = (f_p + g_p \pm h_p)E_s(\omega)E_s(\omega) \tag{4.29}$$

$$P_s(2\omega) = (f_s \pm h_s)E_s(\omega)E_s(\omega) \tag{4.30}$$

which correspond to so-called linear difference effects in SHG (SHG-LD). This effect has no analogue in linear optics.

4.1.3 *Nonlinear optical activity and magnetic-dipole contributions (Verbiest et al. 1999)*

Although it is possible to explain nonlinear optical activity within the electric-dipole approximation (i.e., by only including electric-dipole contributions to the nonlinearity of the medium), experimental results indicate that, depending on the material system, magnetic contributions are nevertheless necessary to accurately explain nonlinear optical activity. In that case, irradiating the sample will induce nonlinear polarization and magnetization, and both will act as sources of second-harmonic radiation. The nonlinear part of the induced dipole moment (polarization), including the magnetic-dipole interactions up to first order, is now given by:

$$\mu_i(2\omega) = \sum_{j,k} \beta_{ijk}^{eee} E_j(\omega)E_k(\omega) + \beta_{ijk}^{eem} E_j(\omega)B_k(\omega) \tag{4.31}$$

$$P_i(2\omega) = \sum_{j,k} \chi_{ijk}^{(2),eee} E_j(\omega)E_k(\omega) + \chi_{ijk}^{(2),eem} E_j(\omega)B_k(\omega) \tag{4.32}$$

The subscripts refer to Cartesian coordinates in the laboratory coordinate system. $E(\omega)$ and $B(\omega)$ are the electric and magnetic induction fields, respectively.

In addition, the nonlinear induced magnetic moment (magnetization) is defined as

$$m_i(2\omega) = \sum_{j,k} \beta_{ijk}^{mee} E_j(\omega)E_k(\omega) \tag{4.33}$$

$$M_i(2\omega) = \sum_{j,k} \chi_{ijk}^{(2),mee} E_j(\omega)E_k(\omega) \tag{4.34}$$

The superscripts e and m in Equations 4.31–4.34 refer to an electric-dipole and magnetic-dipole interaction, as previously introduced. The energy diagrams for the tensors $\chi^{(2),eee}$, $\chi^{(2),eem}$, and $\chi^{(2),mee}$ are given in Figure 4.8. Note also that we have assumed in Equations that the higher

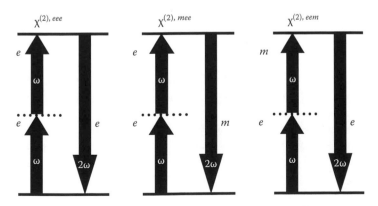

Figure 4.8 Energy diagrams for the tensors $\chi^{(2),eee}$, $\chi^{(2),mee}$, and $\chi^{(2),eem}$.

multipole contributions have predominantly a magnetic-dipole character. In principle, however, one could also include electric-quadrupole interactions (Kauranen et al. 1998). However, because the presence of magnetic-dipole interactions in chiral media is well established, we do not include the quadrupole interactions. In any case, it is experimentally difficult to separate both contributions. One can, however, implicitly include the quadrupole interaction in the magnetic-dipole interaction (see also Chapter 1), which will simplify the discussion considerably.

The non-vanishing components of the tensors $\chi^{(2),eem}$ and $\chi^{(2),mee}$ can be determined by applying the symmetry elements of the medium to the respective tensors. However, to do so one must realize that there is a fundamental difference between the electric field vector and the magnetic field vector. The first one is a polar vector, whereas the latter is an axial vector. A polar vector transforms as the position vector for all spatial transformations, whereas an axial vector transforms as the position vector for rotations and opposite to the position vector for reflections and inversions. Therefore, electric quantities and magnetic quantities transform similarly under rotations, but differently under reflections and inversions. As a consequence, the non-vanishing tensor components of $\chi^{(2),eem}$ and $\chi^{(2),mee}$ can be different from those of $\chi^{(2),eee}$.

Let us, for example, consider the influence of a mirror plane in the yz-plane. The coordinates transform as:

$$x \to -x \qquad y \to y \qquad z \to z$$

Consequently, the electric quantities will transform as:

$$E_x \to -E_x \qquad E_y \to E_y \qquad E_z \to E_z$$
$$P_x \to -P_x \qquad P_y \to P_y \qquad P_z \to P_z$$

However, the magnetic quantities will transform differently:

$$B_x \to B_x \qquad B_y \to -B_y \qquad B_z \to -B_z$$
$$M_x \to M_x \qquad M_y \to -M_y \qquad M_z \to -M_z$$

For example, the mirror plane will transform $M_z = \chi_{zzz}^{mee} E_z E_z$ into $-M_z = \chi_{zzz}^{mee} E_z E_z$, and as a consequence χ_{zzz}^{mee} will be zero for an achiral isotropic surface.

All non-vanishing components of $\chi^{(2),eem}$, $\chi^{(2),mee}$, and $\chi^{(2),eee}$ for films with C_∞ and $C_{\infty v}$ symmetry are given in Table 4.3.

4.1.4   *SHG from chiral films: basic theory*

We start by considering an experimental configuration that is schematically shown in Figure 4.9. A laser beam at frequency ω is incident on a

Table 4.3 Non-Vanishing Components of the Second-Order Susceptibility Tensors $\chi^{(2),eee}$, $\chi^{(2),eem}$, and $\chi^{(2),mee}$ for Chiral and Achiral Films or Surfaces for the case of Second-Harmonic Generation

Tensor	Achiral $C_{\infty v}$ Symmetry	Chiral C_{∞} Symmetry
$\chi^{(2),eee}$	zzz	zzz
	$zxx = zyy$	$zxx = zyy$
	$xxz = xzx = yyz = yzy$	$xxz = xzx = yyz = yzy$
		$xyz = xzy = -yzx = -yzx$
$\chi^{(2),eem}$		zzz
		$zxx = zyy$
		$xxz = yyz$
		$xzx = yzy$
	$xyz = -yxz$	$xyz = -yxz$
	$zxy = -zyx$	$zxy = -zyx$
	$xzy = -yzx$	$xzy = -yzx$
$\chi^{(2),mee}$		zzz
		$zxx = zyy$
		$xxz = xzx = yyz = yzy$
	$xyz = xzy = -yxz = -yzx$	$xyz = xzy = -yxz = -yzx$

Note: The film is taken to be in the x-y plane.

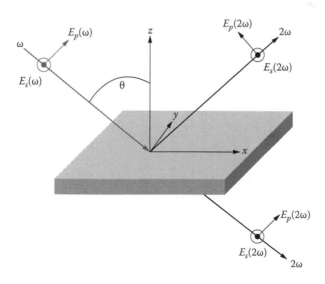

Figure 4.9 Geometry of second-harmonic generation from a thin film or surface.

chiral sample (film or surface) with C_∞ symmetry (in-plane isotropy). The symmetry of such a sample is broken in the z-direction but arbitrary rotations about the surface normal remain symmetry operations.

For an achiral sample ($C_{\infty v}$), reflection in a vertical plane normal to the surface is an additional symmetry operation. The fundamental beam can be expressed in terms of the p- and s-polarized components $E_p(\omega)$ and $E_s(\omega)$ as

$$E(\omega) = pE_p(\omega) + sE_s(\omega) \qquad (4.35)$$

where p and s are unit vectors in the spk coordinate system. They are related to the Cartesian (x,y,z) vectors by

$$\vec{s} = -\vec{y}, \quad \vec{p} = \vec{x}\cos\theta + \vec{z}\sin\theta \qquad (4.36)$$

where θ is the angle of incidence. It is important that we assume that the nonlinear layer is sufficiently thin such that its linear optical activity does not appreciably mix the p- and s-polarized components of the fundamental or second-harmonic fields.

As we have seen earlier, the nonlinear interaction of the fundamental field with the material can be described by the nonlinear polarization and magnetization up to the first order in the magnetic-dipole interaction:

$$P_i(2\omega) = \sum_{j,k} \chi_{ijk}^{eee} E_j E_k + \chi_{ijk}^{eem} E_j B_k \qquad (4.37)$$

$$M_i(2\omega) = \sum_{j,k} \chi_{ijk}^{mee} E_j E_k \qquad (4.38)$$

The nonlinear polarization $P(2\omega)$ and magnetization $M(2\omega)$ act as sources of second-harmonic radiation and the electric-field amplitude of the second-harmonic generation is directly proportional to the two sources. In addition, the magnetic field $B(\omega)$ depends linearly on the electric field $E(\omega)$. As a consequence, we can write the second-harmonic generated field as (Kauranen et al. 1994):

$$E_i(2\omega) = f_i E_p^2(\omega) + g_i E_s^2(\omega) + h_i E_p(\omega) E_s(\omega) \qquad (4.39)$$

where f_i, g_i, and h_i are linear combinations of the components of the nonlinear susceptibilities χ^{eee}, χ^{eem}, and χ^{mee}. If we neglect linear optical parameters, these coefficients are similar to those introduced in Section 4.1.2. However, here they have a more general meaning: the detailed forms of the coefficients are different depending on which contributions are included in the theory. For example, one can include the relevant linear optical parameters of the sample and surrounding media for a complete description of the optical response. In fact, Maki et al. (1995) used a more detailed form of *f*, *g*, and *h* that also included linear and nonlinear Fresnel factors to describe the nonlinear optical response of chiral thin films. Furthermore, one could also include electric-quadrupole interactions.

The exact form of the coefficients *f*, *g*, and *h* is also dependent on the symmetry of the sample. For a film or surface with C_∞, the explicit expressions for such surfaces are (if we neglect linear optical parameters):

$$f_{Ts}^{Rs} = -\sin\theta \left(-2\chi_{xyz}^{eee}\cos\theta - \chi_{xzx}^{eem} + \chi_{zzz}^{mee}\sin^2\theta \right.$$
$$\left. +\chi_{zxx}^{mee}\cos^2\theta \mp 2\chi_{xxz}^{mee}\cos^2\theta \right) \tag{4.40}$$

$$g_{Ts}^{Rs} = -\sin\theta \left(\chi_{xxz}^{eem} + \chi_{zxx}^{mee} \right) \tag{4.41}$$

$$h_{Ts}^{Rs} = \sin\theta \left(2\chi_{xxz}^{eee} - \left(\chi_{xzy}^{eem} + \chi_{xyz}^{eem} \right)\cos\theta \mp 2\chi_{xyz}^{mee}\cos\theta \right) \tag{4.42}$$

$$f_{Tp}^{Rp} = \sin\theta \left(\chi_{zzz}^{eee}\sin^2\theta + \chi_{zxx}^{eee}\cos^2\theta \mp 2\chi_{xxz}^{eee}\cos^2\theta \right.$$
$$\left. -\chi_{zxy}^{eem}\cos\theta \pm \chi_{xzy}^{eem}\cos\theta + 2\chi_{xyz}^{mee}\cos\theta \right) \tag{4.43}$$

$$g_{Tp}^{Rp} = \sin\theta \left(\chi_{zxx}^{eee} - \chi_{zxy}^{eem}\cos\theta \mp \chi_{xyz}^{eem}\cos\theta \right) \tag{4.44}$$

$$h_{Tp}^{Rp} = -\sin\theta \left(\mp 2\chi_{xyz}^{eee}\cos\theta + \left(\chi_{zzz}^{eem} - \chi_{zxx}^{eem} \right)\sin^2\theta \right.$$
$$\left. \mp \left(\chi_{xzx}^{eem} + \chi_{xxz}^{eem} \right)\cos^2\theta - 2\chi_{xxz}^{mee} \right) \tag{4.45}$$

where *R* and *T* refer to the second-harmonic fields generated in the reflection and transmission direction, respectively; *s* and *p* refer to the polarization of the detected signal; and *θ* is the angle of incidence. It is clear that for an achiral sample and for the *s*-polarized second-harmonic component, only h_s is non-vanishing (because h_s is composed of achiral susceptibility components), while for a chiral sample f_s, g_s and h_s are non-vanishing (f_s and g_s are therefore composed of chiral susceptibility components). Note also that the coefficient g_s vanishes in the electric-dipole approximation. If we analyze the equations for *p*-polarized detection, we can see that the main difference is that *f* and *g* are in this case composed of achiral susceptibility components, while *h* only contains chiral components.

We will consider throughout this chapter *s*- and *p*-polarized detection. This is important because the role of *f*, *g*, and *h* as chiral or achiral components is reversed between the *s*- and *p*-polarized second-harmonic signals. Therefore, any configuration that mixes p and s polarizations, gives rise to nonzero values of the three coefficients, even though the sample is achiral.

4.1.4.1   Second-harmonic generation—Circular dichroism

For the case of circularly polarized fundamental beams, $E_p(\omega) = \pm iE_s(\omega)$, the intensity of the second-harmonic field is found from Equation 4.39 and $I \propto |E|^2$ to be:

$$I(2\omega) \propto |-f + g \pm ih|^2 \, |E_s(\omega)|^4 \tag{4.46}$$

The circular-difference response is arising from the interference between the quantities g-f and h. Since f_s and g_s and h_p are zero for an achiral sample, no circular-difference effect can occur for p-polarized or s-polarized detection. Hence, SHG-CD will only occur for chiral materials and the effect will change the sign for the two enantiomers of the sample. Also for a CD-effect to occur, there has to be a phase difference between g-f and h.

Within the electric-dipole approximation, this phase difference will exist if the fundamental and/or second-harmonic frequencies are close to resonance frequencies of the material. On the other hand, when we include magnetic-dipole interactions, the required phase difference will also exist for nonresonant excitation. Note that the two conditions have to be fulfilled simultaneously in order to observe SHG-CD effects. These conditions are the presence of chiral and achiral expansion coefficients and a phase difference between these coefficients. Therefore, the lack of an SHG-CD effect cannot be used to conclude that the material is achiral.

The difference in SH intensity for left- and right-hand circularly polarized excitation $\Delta I(2\omega)$, is given by

$$\Delta I(2\omega) = I(2\omega)_{left} - I(2\omega)_{right} = 4\,\mathrm{Im}((f - g)h^*)I(\omega)^2 \tag{4.47}$$

In general, the SHG-CD effect can be described by a normalized quantity (as in linear optics $\Delta\varepsilon/\varepsilon$):

$$\frac{\Delta I(2\omega)}{I(2\omega)} = \frac{I_{LCH}(2\omega) - I_{RCH}(2\omega)}{|I_{LCH}(2\omega) + I_{RCH}(2\omega)|/2} \tag{4.48}$$

We can also use linear input polarizations to probe surface chirality. For the case of linear polarizations that are rotated by $\pm 45°$ from the

p-polarized direction, $E_p = \pm E_s$ and the intensity of the second-harmonic field is

$$I(2\omega) \propto |f + g \pm h|^2 |E_s(\omega)|^4 \tag{4.49}$$

Hence, a different SH-efficiency can also occur for ±45° linearly polarized light and this effect is known as second-harmonic generation-linear dichroism (SHG-LD). In order to observe SHG-LD, $f + g$ and h must be non-vanishing (the material has to be chiral) and $f + g$ and h have to be in-phase. The linear difference response is given by:

$$\Delta I(2\omega)_{-45} - \Delta I(2\omega)_{+45} = 4\,\mathrm{Re}((f + g)h^*)I(\omega)^2 \tag{4.50}$$

The difference can therefore exist in nonresonant excitation in the electric-dipole approximation. In this respect, linear-difference effects are complementary to circular-difference effects in SHG (where a phase difference between the expansion coefficients is required). This linear-difference response has no analogue in linear optics.

4.1.4.2 SHG-ORD

Second-harmonic generation optical rotatory dispersion (SHG-ORD) can be considered the nonlinear analogue of linear optical rotation. While SHG-CD requires complex susceptibility components, that is, it is a resonance effect, SHG-ORD can also occur under nonresonant conditions. Hence, SHG-ORD can be considered as a surface-sensitive and chiral-selective technique that does not suffer from the constraint of resonance.

SHG-ORD uses linearly polarized fundamental light and the polarization maximum of the outgoing SHG signal is observed as a function of the rotation angle ϕ. The observed SHG-ORD will be characterized by this angle, which is the amount of rotation of the SHG signal in comparison with an achiral surface. Out of resonance, ϕ can be defined as

$$\phi = \tan^{-1}[R(2\omega)] \tag{4.51}$$

where $R(2\omega)$ is the ratio of the s-polarized field component to the p-polarized component:

$$R(2\omega) = \frac{E_s(2\omega)}{E_p(2\omega)} = \frac{f_s E_p^2(\omega) + g_s E_s^2(\omega) + h_s E_p(\omega)E_s(\omega)}{f_p E_p^2(\omega) + g_p E_s^2(\omega) + h_p E_p(\omega)E_s(\omega)} \tag{4.52}$$

However, this equation can be problematic in the sense that an achiral sample can also lead to polarization rotation, because h_s, f_p, and g_p are nonzero for achiral systems. This problem is solved by always using p-polarized fundamental light. In that case, the expression is considerably simplified, and optical rotation is only observed for chiral materials:

$$\phi = \tan^{-1}\left(\frac{f_s}{f_p}\right) \qquad (4.53)$$

The angle ϕ is dependent upon the chirality of the sample. The chirality of the sample is reflected through the sign and magnitude of f_s. As we know, the sign of f_s reverses between two enantiomers, and the sign of the achiral coefficient f_p does not change. Therefore, ϕ will change the sign between the two enantiomers. As a consequence, different enantiomers of a chiral molecule can be distinguished by the sign of the SHG-ORD rotation angle.

The magnitude of the SHG-ORD effect will depend on the ratio of chiral and achiral components (Figure 4.10). ϕ will be large if f_s is large compared to f_p. No resonance is required for SHG-ORD to occur, although f_s can be resonantly enhanced, which would increase the ratio in equation 4.53. Note also that the rotation angle is limited to ±90°, while in linear

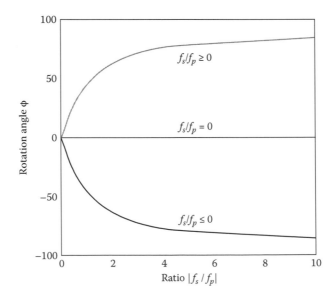

Figure 4.10 Rotation angle of the polarization of the second-harmonic light versus the absolute value of the ratio f_s/f_p.

optics optical rotation can add up unlimitedly by propagating through the chiral sample.

4.1.5   Sum-frequency generation in chiral isotropic liquids (Ji and Shen 2006)

In sum-frequency generation, the pulses from two lasers are overlapped in a medium, and light is generated at the sum of the two incident beams. The effect has been used to study thin films and surfaces, especially to probe vibrational transitions. The nonlinear polarization can be described in the electric-dipole approximation as:

$$P_i(\omega_1+\omega_2)=2\sum_{j,k}\chi^{(2)}_{ijk}E_j(\omega_1)E_k(\omega_2)$$

(4.54)

Similar symmetry considerations apply as in the case of SHG, except for the nondegeneracy of the two input frequencies. Hence, SFG is—at least within the electric-dipole approximation—forbidden in centrosymmetric systems such as liquids. However, chiral liquids or solutions of chiral molecules are inherently noncentrosymmetric, and one could expect a second-order nonlinear response. If we apply the symmetry of a chiral liquid to the susceptibility tensor we end up with six non-vanishing components:

$$\chi^{(2)}_{xyz}=\chi^{(2)}_{yzx}=\chi^{(2)}_{zxy}=-\chi^{(2)}_{xzy}=-\chi^{(2)}_{yxz}=-\chi^{(2)}_{zyx}$$

(4.55)

It is immediately clear that these components would vanish for the case of SHG due to the degeneracy of the two input intensities. The components of the nonlinear polarization can then be written as:

$$P_x(\omega_1+\omega_2)=2\chi^{(2)}_{xyz}E_y(\omega_1)E_z(\omega_2)+2\chi^{(2)}_{xzy}E_z(\omega_1)E_y(\omega_2)$$

$$P_y(\omega_1+\omega_2)=2\chi^{(2)}_{yxz}E_x(\omega_1)E_z(\omega_2)+2\chi^{(2)}_{yzx}E_z(\omega_1)E_x(\omega_2)$$

$$P_z(\omega_1+\omega_2)=2\chi^{(2)}_{zyx}E_y(\omega_1)E_x(\omega_2)+2\chi^{(2)}_{zxy}E_x(\omega_1)E_y(\omega_2)$$

(4.56)

or in vector notation:

$$\vec{P}(\omega_1+\omega_2)=2\chi^{(2)}(\vec{E}(\omega_1)\times\vec{E}(\omega_2))$$

(4.57)

It follows from these equations that the electric fields at frequency ω_1, ω_2, and $\omega_1+\omega_2$ need to span the x, y, and z directions of a Cartesian frame. As a consequence, this requires a noncolinear beam geometry with one *s*-polarized and two *p*-polarized beams. Whereas SHG-CD probes chirality with circularly polarized light, SFG does not require circularly polarized light. Instead it is the detection of photons at the sum frequency that constitutes the chiral measurement. As a consequence, SFG in solutions cannot distinguish between different enantiomers, but can only probe for the presence of chirality.

The intensity of the SFG process is rather weak, and in practice at least one of the frequencies should be near or on resonance to obtain a measurable signal. Sum-frequency generation can probe electronic as well as vibrational transitions. If one of the input beams is tunable in the infrared, SFG may be used to record the vibrational spectrum of a chiral molecule.

4.2   Experimental procedures

4.2.1   Continuous polarization measurements

In the previous section, we introduced SHG-CD, SHG-LD, and SHG-ORD. Although these techniques are useful for the detection of surface chirality, it is difficult to differentiate between the enantiomers. However, all structural information is embedded in the nonlinear susceptibility tensors. Hence, a complete analysis of all components of the susceptibility tensor would be useful. A method that is able to do so is based on a series of continuous polarization measurements and is a simplified version of the procedure outlined in Section 3.2.2 in the sense that only a quarter-wave plate is used instead of a combination of a half- and quarter-wave plate. Here, the polarization of the fundamental is continuously varied by a rotating quarter-wave plate. The intensity of the second-harmonic light is recorded in transmission and reflection as a function of the rotation angle of the quarter-wave plate. Such measurements are referred to as "polarization measurements." The polarization patterns contain direct information on the circular-difference response and on the chirality of the sample.

The p- and s-polarized components of the fundamental beam can be expressed as:

$$E_p(\omega) = E_0(\cos^2 \varphi + i \sin^2 \varphi) = P(\varphi)$$

(4.58)

$$E_s(\omega) = E_0 \sin \varphi \cos \varphi (1 - i) = S(\varphi)$$

(4.59)

where E_0 is the amplitude of the initial fundamental field, φ is the rotation angle between the p-polarized direction and the axis of the wave plate, and $P(\varphi)$ and $S(\varphi)$ represent the functional dependence of the fundamental-field components on the rotation angle. The intensity of the second-harmonic field then has the general form:

$$I(\varphi) \propto |E(2\omega)|^2 = |fP^2(\varphi) + gS^2(\varphi) + hP(\varphi)S(\varphi)|^2$$

(4.60)

When the detected signal is s-polarized and the surface or film is achiral ($C_{\infty v}$), the only non-vanishing expansion coefficient is h_s. The polarization pattern of the second-harmonic intensity is then described by:

$$I_s(\varphi) \propto |h_s^2| |P(\varphi)S(\varphi)|^2 \tag{4.61}$$

This equation is an even function of the rotation angle φ and therefore the polarization pattern will be symmetric around the 180° rotation angle of the quarter-wave plate. No circular-difference effects will be present (Figure 4.11a). For a chiral sample, on the other hand, the polarization pattern is described by Equation 4.60. Depending on the relative phase of the coefficients f_s, g_s, and h_s, the polarization line shapes can have a very different form but there will always be an asymmetry around the 180° rotation angle (Figure 4.11b and 4.11c). However, circular-difference effects will only be present when there is a phase difference between the expansion coefficients (Figure 4.11c).

Similar arguments hold for the p-polarized signals. For an achiral sample, the p-polarized second-harmonic intensity is:

$$I_p(\varphi) \propto |f_p P^2(\varphi) + g_p S^2(\varphi)|^2 \tag{4.62}$$

which is an even function of the rotation angle φ because $P^2(\varphi)$ and $S^2(\varphi)$ are even functions of φ (Figure 4.11d). On the other hand, for the case of a chiral sample, the intensity has the general form of Equation 4.60. Hence, we see again that chirality leads to a left-right asymmetry in the polarization pattern $I_p(\varphi)$ (Figure 4.11e and 4.11f). This asymmetry is reversed if the handedness of the sample is reversed. Similarly, we will only observe CD effects if there is a phase difference between the components.

In general, the lack of mirror symmetry with respect to the 180° rotation angle φ in the p- or s-polarized signals is an evidence of chirality. While for SHG-CD and SHG-LD effects certain requirements of resonance are to be fulfilled to detect the chirality of a sample, left-right asymmetry in the polarization pattern is a direct evidence of chirality. The relative phase of the achiral and chiral components will only change the details of the asymmetry.

Quantitative information can be obtained from these patterns by fitting them to the following equation:

$$I(2\omega) = \frac{1}{16}[f_R - g_R + 4f_I \cos 2\varphi - (f_R - g_R)\cos 4\varphi + 2h_I \sin 2\varphi - h_R \sin 4\varphi]^2 I^2(\omega)$$

$$+ \frac{1}{16}[f_I - g_I - 4f_R \cos 2\varphi - (f_I - g_I)\cos 4\varphi - 2h_R \sin 2\varphi - h_I \sin 4\varphi]^2 I^2(\omega)$$

$$\tag{4.63}$$

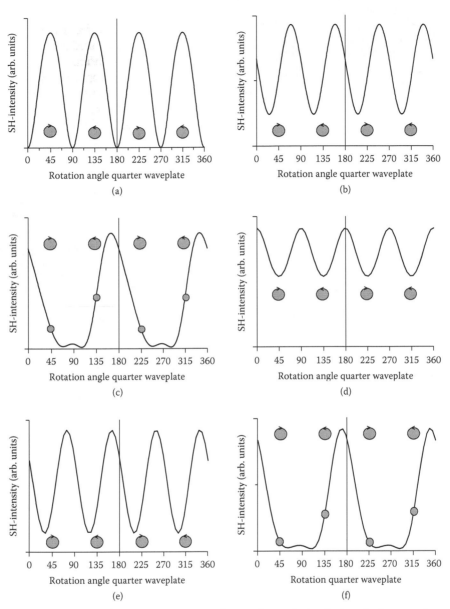

Figure 4.11 (a) Polarization patterns of *s*- and *p*-polarized signals for quarter-wave plate measurements for (a) an achiral sample ($C_{\infty v}$), *s*-polarized second-harmonic; (b) a chiral sample (C_{∞}), *s*-polarized second-harmonic, with both f_s and h_s real; (c) a chiral sample (C_{∞}), *s*-polarized second-harmonic, with both f_s complex and h_s real; (d) an achiral sample ($C_{\infty v}$), *p*-polarized second-harmonic; (e) a chiral sample (C_{∞}), *p*-polarized second-harmonic, with f_p, g_p, and h_p real; and (f) a chiral sample (C_{∞}), *p*-polarized second-harmonic, with f_p, g_p complex and h_p real. Circular difference effects are indicated by gray circles.

where f_R, g_R, h_R, f_I, g_I, and h_I are the real and imaginary parts of the coefficients f, g, and, h respectively. As mentioned before, for a complete description, linear refractive indices can be included in the coefficients. From this
equation, one can extract the values of f, g, and h. To fit the experimental
results, it is necessary to fix the overall phase. This can be done, for example, by defining h as a real quantity ($h_I = 0$). The values found for the coefficients f, g, and h can then subsequently be used to calculate the values of
the components of the second-order susceptibility, $\chi^{(2)}$. Note also that only
the relative phase and magnitude of all tensor components can be determined. This is due to the arbitrary choice of $h_I = 0$. In addition, all second-
harmonic intensities are recorded in arbitrary units, so that no absolute
magnitude of a component can be calculated. However, absolute values of
a component can be determined by calibration with a quartz wedge.

A convenient system that allows for very accurate polarization measurements at a single wavelength is based on a Nd:YAG laser as the light
source (Figure 4.12). In a typical system, the fundamental beam of a Nd:
YAG laser is sent through a combination of a half-wave plate and a polarizer, which allows adjustment of the input intensity. Next, the input beam
is p-polarized by a high quality polarizer and subsequently focused
(typical focal length 20–50 cm) on the sample at an incidence angle of
45°. The polarization of the fundamental beam is continuously varied by
a rotating quarter-wave plate and the intensity of the second-harmonic
(s- or p-polarized) is recorded in reflection and transmission. The light
of the flash lamps of the laser are isolated from the experiment by a
visible-blocking filter. The reflected and transmitted second-harmonic

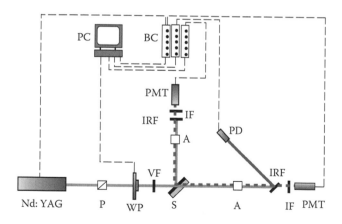

Figure 4.12 Experimental setup for continuous polarization measurements. P =
polarizer, WP = wave plate, VF = visible light filter, S = sample, A = analyzer, IRF =
infra filter, IF = interference filter, PMT = photomultiplier tube, PD = photodiode,
BC = boxcar integrator.

beams are passed through infrared blocking and a 532-nm interference filter before detection by a photomultiplier tube (PMT). The polarization of the fundamental beam is continuously monitored by a photodiode that detects reflected light from the infrared (IR) blocking filter. The data acquisition is done by a boxcar gated integrator and a computer. Important is the use of good polarizing optics; the slightest polarization impurity can lead to erroneous signals that can seriously affect the data analysis.

4.2.2   Probing the chiral xyz component to study chirality

The simplest way to study chirality in samples with in-plane isotropy is to monitor the chiral susceptibility components of the *xyz* (electric-dipole) type. As is evident from Equations 4.39–4.45, this can be easily done by measuring the *s*-polarized second-harmonic component for a *p*-polarized fundamental beam. In the absence of magnetic-dipole terms, the only component that is being probed is the $\chi^{(2)}_{xyz}$-component. If additional magnetic-dipole terms are present, they also contribute to the signal. In any case, the presence of a nonzero *p*-in *s*-out signal for a sample with C_∞ is direct proof of chirality. Hence, by measuring the wavelength dispersion of this component, one is able to measure a nonlinear analogue of the classical CD spectrum. An important difference, however, is that one is not able to distinguish between the enantiomers. Indeed the measured signal is proportional to $\left|\chi^{(2)}_{xyz}\right|^2$ and any information concerning the sign of this component is lost. In principle, this problem could be addressed by interfering the signal with a reference signal with known phase.

4.2.3   SHG-CD and SHG-ORD measurements

SHG-CD and SHG-ORD measurements require a tunable laser system with a relatively high peak power. There are several options available depending on the wavelength region of interest. One option is a tunable, amplified femtosecond system comprising four distinct lasers. These include a seed laser and a pump laser. The seed pulses (800 nm) are amplified in a regenerative amplifier to produce femtosecond pulses with a high energy per pulse. Then the light is sent to an optical parametric amplifier (OPA) that changes the wavelength from any specific value between approximately 1100 and 1600 nm.

   A typical experimental setup can be described as follows. After leaving the OPA, the beam is sent through a combination of a half-wave plate and a polarizer to adjust the intensity. Next, a Berek compensator is used to change the input polarization. Further on, a lens (typical focal length of 20 cm) focuses the beam onto the sample through a visible light filter that blocks the visible part of the spectrum while transmitting 90% of the infrared radiation. The sample holder is typically mounted on a rotation stage permitting the variation of the angle of incidence with respect to the sample normal. Afterward, an analyzer enables one to select the different components of the second-harmonic radiation. A subsequent IR filter and monochromator ensure the detection of the proper wavelength by a PMT, which is connected to a gated integrator. Two IR photodiodes are used in the setup to monitor the IR intensity. The first one is placed at the output of the OPA to follow the stability of the laser and if necessary to correct laser intensity drift in the obtained results. The second photodiode detects the reflection of the IR filter and monitors the polarization of the fundamental light as a reference. It can also reveal the occurrence of sample damage.

The data acquisition is done by the combination of a gated boxcar integrator and a PC. The boxcar integrator is triggered by the output signal of the amplifier, which occurs at the moment the laser pulse leaves the amplifier. To improve the signal-to-noise ration, it is convenient to use the "active baseline subtraction" feature of the boxcar integrator, in which every second detected electronic pulse is inverted and added to an average. This can be done by blocking every second laser pulse so that the boxcar then measures only ambient noise and effectively subtracts it from the signal. This practice is known to improve the signal-to-noise ratio by an order of magnitude at least.

Finally, care has to be taken to ensure good polarization purity. Therefore, high quality polarizers combined with an adjustable Berek compensator is absolutely necessary.

Another option to measure SHG-CD (or SHG-ORD) is to make use of the wavelength tunability of the Ti-sapphire system. In principle this can be tuned in a wavelength region between 700 and 900 nm. The experimental setup is similar to the one described above, but the detection scheme is different. For all practical purposes, the 80 MHz pulse frequency of the Ti-sapphire system can be treated as quasi-continuous. Therefore, phase-sensitive detection techniques can be used to analyze the SH light. Hence, the laser beam is chopped at approximately 1 kHz and the PMT is connected to the high impedance input of a lock-in detector. The high impedance raises the sensitivity of the measurement, while dark current and other background signals are eliminated by the proper action of phase-sensitive detection.

4.3   Second-harmonic generation in nanostructures (Canfield, Kujala, Jefimous et al. 2006)

A formalism very similar to that described in Section 4.2.1 has been developed to describe the nonlinear optical response of metal nanostructures. Describing the second-order response from nanostructures is not an easy task because rapid variations in local symmetry due to defects, plasmonic interactions, and local field contributions make it practically impossible to establish a relationship between the macroscopic response and the nanoscopic description of the nonlinearities. Furthermore, higher multipole interactions may be present and complicate further analysis. However, by using a macroscopic nonlinear response tensor (NRT) approach, in which the sample is basically treated as a "black box," one is able to circumvent these problems. The procedure is based on the fact that the field at the second-harmonic frequency is of the general form

$$E_j(2\omega) = f_j E_p^2(\omega) + g_j E_s^2(\omega) + h_j E_p(\omega)E_s(\omega) \tag{4.64}$$

The meaning of the coefficients f, g, and h is very similar to the coefficients used in Section 4.1. Here, they can be considered as quantities that characterize the nonlinear optical response, without making any assumptions about the nature of the nonlinearity. While in the previous sections we have connected these coefficients to the different nonlinear sources through the respective susceptibilities, in this case the coefficients are simply considered as quantities that relate the input and the output fields without any further details. However, the problem with this equation is that the natural coordinate system of the fields (*spk*) is different from that of the sample (*xyz*). Hence, it will in general be very difficult to apply any rules related to the sample symmetry. However, in the case that the sample and experiment are performed in the same geometry, the situation becomes much simpler. For example, for an experimental configuration where the light is incident under normal incidence, both the sample and field can be treated in the *xyz* coordinate system. The second-harmonic field is than of the form:

$$E_j(2\omega) = A_{jxx}E_x^2(\omega) + A_{jyy}E_y^2(\omega) + A_{jxy}E_x(\omega)E_y(\omega) \tag{4.65}$$

where the tensor A is the macroscopic response tensor. In this particular experimental situation, the classical electric-dipole symmetry rules (see Section 1.4) can be applied to the sample, and, as a consequence, valuable information concerning symmetry breaking and defects in the particles can be observed. For example, regular cylindrical arrays of gold nanoparticles were recently investigated. Although SHG is forbidden for this particular arrangement of particles, SHG was observed indicating the presence of structural defects in the nanoparticles (Canfield et al. 2008). Even strong chiral effects in SHG have been observed in nanoparticles due to inhomogeneities in the samples (Canfield, Kujala, Laiho et al. 2006).

Hence, by applying this approach for the study of nanoparticles it is possible to gain valuable information on the quality of the nanoparticles.

4.4   Applications to biological systems

Chiral-surface SHG has been applied to biological molecules and has recently matured into a powerful characterization technique in this field. Initially, SHG-CD focused on relatively simple biological systems. For example, one of the first studies on biological molecules was on the dipeptide (*ter*-butyloxycarbonyl-tryptophan-tryptophan) on an air–water interface. SHG-CD values of up to 150% were found and it was shown that the DD form of the Boc-Trp-Trp gave a SHG-CD response of equal magnitude but opposite sign, compared with the response of the LL form of the same molecule. The 50/50 racemic mixture gave no SHG-CD response. Hence, it was demonstrated that the handedness of the dipeptide conformation could be revealed by SHG-CD (Crawford et al. 1994). Another example is the study of the redox membrane protein cytochrome c, where it was shown that SHG-CD is sensitive to the heme redox state, which is central to the function of the protein (Petralli-Mallow at al. 2000). However, probably the most important application of SHG-CD is for the study of protein adsorption and protein dynamics. This can be easily done by probing the chiral susceptibility component of the surface. Since proteins are inherently chiral, this component can be used as a spectroscopic probe to measure protein adsorption or protein dynamics. Probing the chiral component can be achieved by selecting an appropriate combination of the polarization of the fundamental beam and the second-harmonic beam. For example, in the case of a sample with in-plane isotropy, this is done by choosing a p-polarized fundamental and the s-polarized second-harmonic component. A more elaborate strategy is to use two input beams in a counter propagating geometry. Certainly for imaging, this strategy has several advantages and was recently demonstrated. It would allow for label-free measurement of protein and DNA binding.

The intrinsic chirality of proteins has been used to study protein binding and to monitor their association to a surface. This is basically done by monitoring the chiral component of the susceptibility upon association. This was recently demonstrated for the binding of mellittin, a hemolytic peptide isolated from bee venom, to lipid bilayers. The importance of this type of experiments lies in the fact that in situ monitoring of surface chirality cannot be studied by conventional methods such as circular dichroism and optical rotatory dispersion due to path length and concentration limitations (Kriech and Conboy 2003).

In principle, binding of a protein to an achiral surface is much simpler than the procedure outlined in Section 3.5. If we use the *p–s* polarization combination to probe the chiral *xyz* components, the second-harmonic intensity (in the electric dipole approximation) versus time is simply given by:

$$I(2\omega) = K \left| \chi^{(2)}_{xyz} \right|^2 I^2(\omega) = K \left| N_s \Theta(t) f\beta \right|^2 I^2(\omega) \tag{4.66}$$

where Θ is as defined in Section 3.5, N_s is the surface density, f is the local field correction, and β is the hyperpolarizability or combination of hyperpolarizability components.

4.5 *Molecular origin*

From a fundamental point of view, a full understanding of the molecular and surface origin of the nonlinear optical activity effects provides a means to relate the experimental measurements back to the molecular structure. However, despite several experimental and theoretical studies, no consensus exists on the molecular and macroscopic origin of these nonlinear optical effects. In general, there are two models used to explain nonlinear optical activity in SHG and SFG. The presence of magnetic-dipole contributions that interfere with electric-dipole contributions are very similar to absorbance circular dichroism. On the other hand a purely electric-dipole approach has also been used to describe the nonlinear optical response. Both models have been successfully applied to several chiral systems: some systems can be described purely by electric-dipole nonlinearities, whereas other systems cannot be described without including higher-order contributions. It is still not clear why in some cases magnetic-dipole contributions have to be included. Attempts have been made to relate both approaches to molecular and supramolecular structures. It has been assumed that the presence of magnetic-dipole contributions may arise from the motion of an electron along a helical path. On the other hand, electric-dipole effects could arise from chirality within the chromophore, such as the presence of a pair of coupled oscillators and they could also arise from achiral chromophores that assume a chiral supramolecular orientation. While this definitively sounds logical, it does not concur with the experimental evidence that has been found during the few years.

Magnetic-dipole contributions have been observed in chiral chromophore-substituted polyisocyanides, a chiral penthamethinium salt, and several chiral conjugated polymers. On the other hand, chiral molecules such as bipnaphtol, helicene derivates, and bacteriorhodopsin have been described within the electric-dipole approximation (Verbiest et al. 1999). We would like to draw particular attention to the chiral conjugated polymers.

More specifically, it has been shown that magnetic-dipole contributions dominate the nonlinear optical response. However, these systems are a perfect example of a coupled oscillator: they stack on top of one another in a chiral fashion. As a consequence, the CD spectrum shows a bisignate Cotton effect in the $\pi-\pi^*$ transition, indicative of chiral exciton coupling. In the helicene derivatives one would expect the electron to move along a helical path giving rise to significant magnetic-dipole contributions. However, these were never observed and the nonlinear optical response can be entirely described within the electric-dipole approximation. Hence, it is clear that it is not straightforward to link molecular or supramolecular structure to the type of nonlinearity.

4.6   Relation with the Faraday effect

Faraday rotation is the rotation of the plane of polarization of linearly polarized light in the presence of a static magnetic field along the light propagation direction. The effect has been known for more than a hundred years and is commonly used in optical isolators, for remote sensing of magnetic fields, or in magneto-optical imaging. Although Faraday rotation may look very similar to optical rotation in chiral materials, the origin of the effect is quite different. Optical rotation in chiral molecules is a linear optical effect, whereas Faraday rotation is in principle a second-order nonlinear optical effect. Furthermore, optical rotation is a reciprocal and Faraday rotation is a nonreciprocal effect. Magnetic circular dichroism, on the other hand, is the differential absorption of left- and right-hand circularly polarized light. Both effects arise from a difference in refractive index for left- and right-hand circularly polarized light. The Faraday rotation angle is given by $\theta = VBL$ with V the Verdet constant, B the component of the magnetic induction field parallel to the light propagation direction, and L the sample length. In terms of its radiation sources, Faraday rotation can be described by a nonlinear polarization:

$$P(\omega) = \chi^{(1),ee}E(\omega) + \chi^{(2),eem}E(\omega)B(0) \tag{4.67}$$

with $B(0)$ the static magnetic field. Important to note is the presence of the $\chi^{(2),eem}$ -component, very similar to those found in chiral materials, except for the frequency dependence. If we have circularly polarized light traveling along the z-direction, $E^{\pm} = E_0(x \pm iy)(e^{ikz-i\omega t} + cc)$, and a static magnetic field along the z-direction, we obtain:

$$P^{\pm}(\omega) = \chi^{(1),ee}E^{\pm}(\omega) \pm i\chi^{(2),eem}E^{\pm}(\omega)B_z(0) = \chi^{\pm}_{eff}E^{\pm}(\omega) \tag{4.68}$$

Using the general equation for the refractive index, we immediately obtain:

$$n^2 = \sqrt{n_0^2 \pm 4\pi i \chi^{(2),eem} B_z(0)} \qquad (4.69)$$

and by using a Taylor expansion:

$$n^{\pm} = n_0 \pm \frac{2\pi i \chi^{(2),eem} B_z(0)}{n_0} \qquad (4.70)$$

The optical rotation θ is given by

$$\theta = \frac{\pi L}{\lambda}(n^+ - n^-) = \frac{4\pi^2 i \chi^{(2),eem}}{\lambda} B_z(0) L \qquad (4.71)$$

Since θ is also equal to VBL, we obtain the relationship between the Verdet constant and $\chi^{(2),eem}$ as:

$$V = \frac{4\pi^2 i \chi^{(2),eem}}{n_0 \lambda} \qquad (4.72)$$

For linearly polarized light that is incident on an isotropic sample (experimental configuration shown in Figure 4.13), the relevant polarization

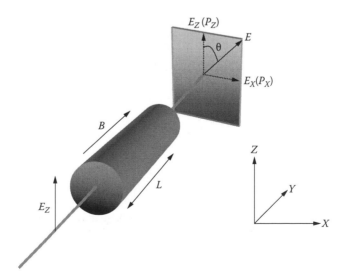

Figure 4.13 Schematic representation of Faraday rotation.

components are:

$$P_z(\omega) = \chi_{zz}^{(1),ee} E_z(\omega)$$

$$P_x(\omega) = \chi_{xzy}^{eem} E_z(\omega) B_y(0)$$

The z-component of the polarization gives rise to a field in the z-direction, while the x-components of the polarization radiate a field in the x-direction. The sum of both fields is a new field that is rotated over an angle with respect to the z-direction. It is immediately clear that the origin of the rotation of polarization lies in the presence of the magnetic-dipole susceptibility component. Note also that a magnetic field in the x- or z-direction would have no influence on the polarization of the light. The reason is that an isotropic sample only contains magnetic susceptibility components of the xyz-type.

References

Byers, J.D., H.I. Yee, and J.M. Hicks, *J. Chem. Phys.*, 101 (1994), 6233.

Canfield, B.K., H. Husu, J. Kontio, J. Viheriälä, T. Rytkönen, T. Niemi, E. Chandler, A. Hrin, J.A. Squier, and M. Kauranen, *New J. Phys.*, 10 (2008), 013001.

Canfield, B.K., S. Kujala, K. Jefimovs, Y. Svirko, J. Turunen, and M. Kauranen, *J. Opt.: Pure Appl. Opt.*, 8 (2006a), S278.

Canfield, B.K., S. Kujala, K. Laiho, K. Jefimovs, J. Turunen, and M. Kauranen, *Opt. Express*, 14 (2006b), 950.

Crawford, M.J., S. Haslam, J.M. Probert, Y.A. Gruzdkov, and J.G. Frey, *Chem. Phys. Lett.*, 229 (1994), 260.

Fisher, P., and F. Hache, *Chirality*, 17 (2005), 421.

Ji, N.A., and Y.R. Shen, *Chirality*, 18 (2006), 146.

Kauranen, M., J.J. Maki, T. Verbiest, and A. Persoons, *J. Chem. Phys.*, 101 (1994), 8193.

Kauranen, M., T. Verbiest, and A. Persoons, *J. Mod. Opt.*, 45 (1998), 403.

Kriech, M.A., and J.C. Conboy, *J. Amer. Chem. Soc.*, 125 (2003), 1148.

Lakhtakia, A., ed., *Selected Papers on Natural Optical Activity*, Bellingham: SPIE, 1990.

Maki, J.J., M. Kauranen, and A. Persoons, *Phys. Rev. B*, 51 (1995), 1425.

Pattanayak, D.N., and J.L. Birman, *Phys. Rev. B*, 24 (1981), 4271.

Petralli-Mallow, T., T.M. Wong, J.D. Byers, H.I. Yee, and J.M. Hicks, *J. Chem. Phys.*, 97 (1993), 1383.

Petralli-Mallow, T.P., A.L. Plant, M.L. Lewis, and J.M. Hicks, *Langmuir*, 16 (2000), 5960.

Shen, Y.R., *The Principles of Nonlinear Optics*, New York: Wiley/Interscience, 1984.

Sioncke, S., T. Verbiest, and A. Persoons, *Mater. Sci. Eng.*, R42 (2003), 115.

Verbiest, T., M. Kauranen, J.J. Maki, M.N. Teerenstra, A.J. Schouten, R.J.M. Nolte, and A. Persoons, *J. Chem. Phys.*, 103 (1995), 8296.

Verbiest, T., M. Kauranen, and A. Persoons, *J. Mater. Chem.*, 9 (1999), 2005.

chapter five

Second-order nonlinear optical imaging techniques

5.1 General principles

5.1.1 Introduction to nonlinear imaging techniques

Imaging of chemical and biological systems based on various spectroscopic techniques is a subject of wide interest. In fact, microscope imaging modes can be based on a wide range of physical principles. There are several types of microscopes: conventional (nonscanning), scanning, and confocal. In fact, microscope imaging modes can be based on any spectroscopy that measures a spatial variation in the signal. Throughout this book, we have seen nonlinear optical processes that are coherent, such as second-harmonic generation; and incoherent, such as hyper-Rayleigh scattering. Hence, the same classification occurs when coupling any of these techniques with a microscope.

Incoherent processes produce a signal whose phase is random and whose intensity is proportional to the concentration of radiating molecules. Fluorescence or so-called spontaneous Raman scattering are common examples of incoherent signals. Fluorescence microscopy has been a powerful technique in cell biology (Kohen and Hirschberg 1989; Pawley 1995), and Raman microscopy has been used in a wide range of applications (Lewis and Edwards 2001). Raman scattering performed in local optical fields of silver and gold nanostructures, the effect of so-called surface enhanced Raman scattering (SERS), can result in effective Raman cross-sections at the level of fluorescence cross-sections (K. Kneipp et al. 1999, 2006; J. Kneipp et al. 2007). The nonlinear equivalent of fluorescence microscopes is based on the simultaneous absorption of two or more photons. The most widely known technique is two-photon excited fluorescence (TPEF) microscopy (Diaspro 2002). Local optical fields also provide the key effect for the observation of surface-enhanced hyper-Raman scattering (SEHRaS) at effective two-photon cross sections similar to or higher than the best cross-sections for two-photon excited fluorescence (J. Kneipp et al. 2007). Hyper-Raman scattering (HRaS) is basically a three-photon process in which a system is excited with two photons of frequency ω and emits one photon at $2\omega \pm \omega_Q$, where ω_Q is the frequency of a vibrational

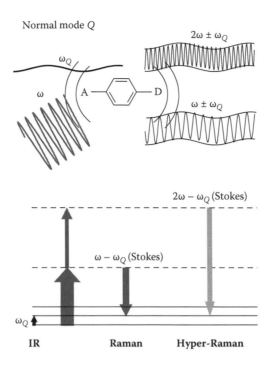

Normal mode Q

$2\omega \pm \omega_Q$

ω_Q

ω

A ———⬡——— D

$\omega \pm \omega_Q$

$2\omega - \omega_Q$ (Stokes)

$\omega - \omega_Q$ (Stokes)

ω_Q

| IR | Raman | Hyper-Raman |

Figure 5.1 Schemes of infrared (IR) absorption, Raman, and hyper-Raman scattering techniques. Raman scattering is a one-photon excitation process; hyper-Raman scattering is the corresponding two-photon excitation process.

excitation (normal mode) in the medium (Figure 5.1). The cross-sections of this nonlinear vibrational spectroscopy are very low. However, a new micro-hyper-Raman setup (Rodriguez et al. 2006) has recently proven to be very efficient even for transparent materials, that is, without local field enhancement or resonance effect, and it also combines a spatial resolution down to the micrometer scale.

Coherent techniques produce signals whose phases are determined by several factors, as we have seen in Chapter 3 for sum-frequency generation (SFG) or second-harmonic generation (SHG), the excitation line, the spatial organization of the radiating molecules, and so forth. In the case of coherent techniques, we have seen that the signal power is now proportional to the square of the concentration of radiating molecules. The coherent nonlinear optical techniques that have the most impact in imaging are SHG and coherent anti-Stokes Raman scattering (CARS) microscopies. CARS spectroscopy has been become one of the well known nonlinear Raman techniques (Clark and Hester 1984), but it is out of the scope of this book since it is a third-order nonlinear optical process. Other coherent techniques such as third-harmonic generation (THG) imaging are the result of work in only the past decade (Barad et al. 1997; Müller et al. 1998), but have benefited also from the fact that it is easier than ever before to generate ultrashort pulses at any desired wavelength.

Two-photon imaging techniques are preferred over single photon methods due to several intrinsic advantages that arise as a result of having a nonlinear intensity-dependent absorption. For instance, because two-photon absorption (which is in fact a third-order process) is restricted to the focal volume, there is in principle no out-of-focus bleaching. In contrast, single-photon absorption and the accompanying bleaching process are quite strong throughout the illuminated volume, as will be detailed in the next section. In addition, because penetration depth is of fundamental importance, two-photon absorption uses excitation wavelengths in the near-infrared range (increased transparency), extending the usable image depth from tens of microns to several hundreds of microns.

5.1.2   Basics of microscopy: Gaussian beam optics

Propagation of Gaussian beams through an optical system can be treated almost as simple geometric optics since a Gaussian beam is transformed into another Gaussian beam (characterized by a different set of parameters). This explains why it is a convenient, widespread model in laser optics. Gaussian beams are usually considered in situations where the beam divergence is relatively small, so that the so-called paraxial approximation can be applied.

For a collimated Gaussian beam propagating through an optical lens, the beam width or spot size will be at a minimum value W_0 at one place along the beam axis (usually at $Z = 0$), known as the beam waist (Figure 5.2).

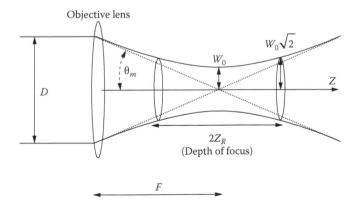

Figure 5.2 Scheme of a collimated Gaussian beam focused through an objective lens.

The parameter W_0, usually called the Gaussian beam radius, is the radius at which the intensity has decreased to $1/e^2$ or 0.135 of its axial value, that is, 86.5% of the encircled power is transmitted. At a distance from the waist equal to the Rayleigh range Z_R, the width (W) of the beam is

$$W(\pm Z_R) = W_0 \sqrt{2} \tag{5.1}$$

where Z_R is defined by

$$Z_R = \frac{\pi W_0^2}{\lambda} \tag{5.2}$$

where λ is the vacuum wavelength divided by the refractive index (n) of the material. The distance between these two points is called the depth of focus (or confocal parameter) of the beam:

$$DOF = 2Z_R = \frac{2\pi W_0^2}{\lambda} \tag{5.3}$$

At large distances from the beam waist (i.e., $Z \gg Z_R$), the beam appears to diverge as a spherical wave from a point source located at the center of the waist. The angle between the asymptotic straight line and the central axis of the beam is called the divergence of the beam and it corresponds to the half-aperture of the microscope lens objective:

$$\theta_m = \frac{\lambda}{\pi W_0} \ (in\ radians) \tag{5.4}$$

In this relation, we have invoked the approximation $\tan(\theta_m) \approx \theta_m$ since the angles are small. Because the origin can be approximated by a point source, θ_m is given by geometrical optics as the diameter illuminated on the lens, D, divided by the focal length, F, of the lens:

$$2\tan(\theta_m) \approx 2\theta_m \approx \frac{D}{F} = \frac{1}{f/\#} \tag{5.5}$$

where $f/\#$ is the photographic f-number (or photographic aperture) of the lens.

Combining Equation 5.4 and Equation 5.5 gives an expression for the beam waist diameter in terms of the input beam parameters and the lens:

$$2W_0 = \frac{4\lambda}{\pi}\left(\frac{F}{D}\right) = \frac{4\lambda}{\pi}(f/\#) \tag{5.6}$$

Therefore, putting Equation 5.6 into Equation 5.3, the depth of focus is now given by the expression

$$DOF = \frac{8\lambda}{\pi}\left(\frac{F}{D}\right)^2 = \frac{8\lambda}{\pi}(f/\#)^2 \tag{5.7}$$

This expression is valid as long as the depth of focus, *DOF*, is less than the focal length, *F*. Another important assumption for all these relations to be valid is that the beam waist occurs at the focal distance from the lens. This is not true for weakly focused beams.

Important is also the numerical aperture, *NA*, of a lens objective, which is defined by the sine of the maximum angle of half-aperture:

$$NA = n\sin(\theta_m) \tag{5.8}$$

where *n* is the refractive index of the medium (usually air). The numerical aperture of a lens determines the ultimate resolution of the optical system. The radial (or lateral) and axial resolution of the objective can be also defined from the numerical aperture of the lens as

$$R_{rad} = \frac{0.61\lambda}{NA}$$

$$Z_{axial} = \frac{DOF}{2} = \frac{2n\lambda}{(NA)^2} \tag{5.9}$$

The radial resolution is limited by diffraction phenomena and corresponds to the radius of the Airy disk. For example, using a high numerical aperture lens of $NA = 0.95$ at a wavelength $\lambda = 632.8$ nm, the visual distinction between two points will take place for a minimum distance of $0.40 \ \mu m$ (Table 5.1).

In the case of tightly focused beams, more sophisticated aspects must be considered since the radius of curvature R of the wavefront evolves according to

$$R(Z) = Z\left[1+\left(\frac{Z_R}{Z}\right)^2\right] \tag{5.10}$$

At the beam waist where $Z = 0$ the phase profile is flat, but along its propagation direction, a Gaussian beam acquires a phase shift that differs from that for a plane wave with the same optical frequency. This difference is called the *Gouy phase shift* (Born and Wolf 1970):

$$\zeta(Z) = -\arctan\left(\frac{Z}{Z_R}\right) \tag{5.11}$$

Table 5.1 Optical Characteristics of Some Objective Lenses (at $\lambda = 632.8$ nm)

Objective Lens Type	50× LWD	50× MPLAN	100× LWD	100× MSPLAN
Half-angle aperture θ_m (in degrees)	33.4	48.6	53.1	71.8
NA $= n \sin(\theta_m)$	0.55	0.75	0.80	0.95
WD (working distance) (mm)	8.1	0.38	3.2	0.37
Photographic f-number: $f/\# = F/D = 1/(2\tan g(\theta_m))$	0.760	0.440	0.375	0.165
Beam waist diameter: $2W_0$ (μm)	0.61	0.35	0.30	0.13
Radial resolution (μm)	0.70	0.51	0.48	0.40
Depth of focus (DOF) (μm)	8.40	4.50	3.95	2.80

It results in a slightly increased distance between wavefronts, compared with the wavelength as defined for a plane wave of the same frequency. This also means that the phase fronts have to propagate somewhat faster, leading to an effectively increased local phase velocity. Overall, the Gouy phase shift of a Gaussian beam for going through a focus (from the far field to the far field on the other side of the focus) is π (Figure 5.3). Additionally, since the Gaussian beam model uses the paraxial approximation, it fails when wavefronts are tilted by more than about 30° from

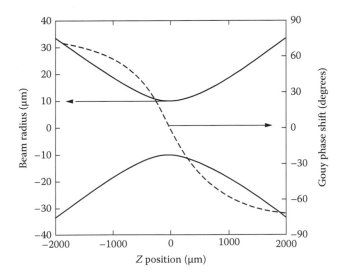

Figure 5.3 Beam radius (solid line) and Gouy phase shift (dashed line) along the propagation direction for a beam in air ($\lambda = 500$ nm, $W_0 = 10$ μm).

the direction of propagation. From the expression for divergence given by Equation 5.4, this means that the Gaussian beam model is strictly valid only for beams with waists larger than about $2\lambda/\pi$. As an illustration, the limiting value at 632.8 nm is $2 \times 0.6328/\pi = 0.40\ \mu m$ and, in Table 5.1, this value corresponds to the best radial resolution obtained with the 100× MSPLAN objective lens type. Then, an objective lens type with higher NA (>1) would lead to more complex relations than those given in this section.

5.1.3 Confocal microscopy

This section gives an overview of the principles of confocal microscopy. Current instruments are highly evolved from the earliest versions, but the principle of confocal imaging advanced by Martin Minsky, and patented in 1957, is employed in all modern confocal microscopes. A diagram of a confocal system is shown in Figure 5.4. The incident laser radiation is focused with the microscope objective on a point in the sample. The backward scattered light from that point is passed back through the objective and focuses on a pinhole aperture. The signal is then picked up by the detector. Signals from points out of the plane containing the focal point is refracted by the objective and passes out of the focus of the pinhole aperture. Consequently, most of the light is blocked by the diaphragm. The major advantage of confocal microscopy is the ability to collect signals from a small volume at the plane of focus (a few or less than cubic micrometers) in a transparent sample, enabling three-dimensional (3-D) imaging by collecting serial optical sections. This optical-sectioning or depth-discrimination property has become the major motivation for using confocal microscopes and is the basis of many of the novel imaging modes of these instruments. However, so far, confocal microscopes in the

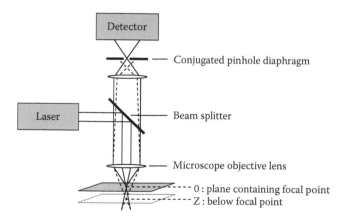

Figure 5.4 Schematic diagram of a confocal microprobe. The solid line represents the depth at which the laser is focused.

linear optical regime are mostly viable in the backward direction, that is, light is currently scattered by fluorescence or Raman processes.

We have seen that the radial resolution is strongly dependent on the wavelength and the numerical aperture of the objective lens (Equation 5.9). The confocal technique enables an optical-sectioning, and limitation in axial resolution is directly connected to the size of the pinhole aperture. Increasing the aperture of the pinhole until full aperture will result in decreasing the axial resolution until it reaches the depth of focus, which is determined by the optical characteristics of the objective lens (Equation 5.7). Indeed, it is necessary to find a compromise between spatial resolution, that is, the ability to distinguish the signal from two adjacent points in the 3-D space, and the quantity of light that is detected. The optimal pinhole diameter for spatial filtering, ϕ_{opt}, is equal to the size of the image given by the microscope objective (magnification) at the focal point in the sample:

$$\phi_{opt} = \gamma \phi_{Airy} \tag{5.12}$$

where γ is the magnificence of the focusing optics augmented by the coupling optics (the coupling factor is usually ~1.4), and ϕ_{Airy} is the diameter of the Airy disk at the focus of the objective, defined by

$$\phi_{Airy} = 2R_{Rad} = 1.22 \frac{\lambda}{NA} \tag{5.13}$$

A few optimal confocal apertures for different types of objective lens are reported in Table 5.2.

In transparent or weakly absorbing media, the laser penetrates inside the sample and one must take into account refraction effects. The optical path is modified according to the Snell–Descartes law (Figure 5.5):

$$n \sin(\theta_m) = n' \sin(\theta'_m) \tag{5.14}$$

where n (n') and θ_m (θ'_m) are the index of refraction and the incidence angle in the first (second) medium.

As a direct consequence, in-depth confocal studies through a planar interface between materials of mismatched refraction indices are affected

Table 5.2 Optimal Pinhole Diameter for Spatial Filtering of Some Objective Lenses (at $\lambda = 632.8$ nm)

Objective Lens Type	50× LWD	50× MPLAN	100× LWD	100× MSPLAN
NA = $n \sin(\theta_m)$	0.55	0.75	0.80	0.95
Radial resolution (μm)	0.70	0.51	0.48	0.40
ϕ_{opt}(μm)	98	71	134	112

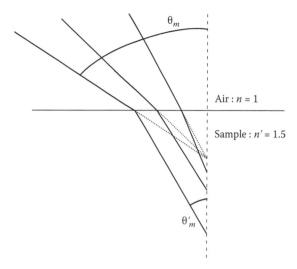

Figure 5.5 Optical paths at a plane interface of a transparent sample with $n' = 1.5$. The marginal (dotted lines in the sample) and paraxial (solid lines in the sample) rays do not converge to the same position and upon increasing the numerical aperture $NA = \sin\theta_m$, the laser irradiance is spread over a larger volume.

by a decrease in both the collected intensity and the axial resolution as a function of focal depth. As illustrated in Figure 5.5, the marginal and paraxial rays do not converge to the same position, giving rise to longitudinal spherical aberration. As a rule of thumb, the internal numerical aperture is lowered and the focus of the beam inside the medium is forwarded in the axial direction. These effects have been well described and characterized by confocal Raman microspectrometry (Everall 2000a, 2000b; Bruneel et al. 2002; Sourisseau 2004). A solution to minimize these effects is to match as much as possible the refraction indices of the two media with an oil/water or solid (Ippolito et al. 2005) immersion objective lens (Figure 5.6).

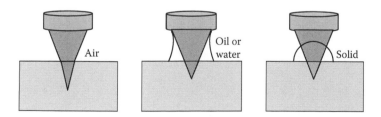

Figure 5.6 In-depth imaging of a transparent sample with a conventional optical microscope (left), an oil or water immersion microscope (middle), and a solid immersion lens microscope (right).

5.2   Experimental techniques and equipment

5.2.1   Two-photon excited fluorescence imaging

Nonlinear microscopy techniques have become important tools for the investigation of biological phenomena where high resolution, 3-D imaging is essential for understanding the underlying biological function (Denk et al. 1990; Svoboda et al. 1997). In TPEF microscopy, two excitations from a pulsed laser combine to excite a fluorescent molecule. The molecule then releases its energy by emitting a fluorescent photon, as is the case for one-photon fluorescence (Figure 5.7). TPEF is, in principle, a third-order nonlinear optical technique, but we will give a brief overview here since it is often used in combination with SHG imaging.

Cross-sections in two-photon excited microscopy are relatively very low and it is mandatory to optimize the experimental conditions to gain a sufficient signal for imaging. In particular, the use of an ultrashort pulse laser source was one of the keys to access nonlinear optical imaging. For input beams with Gaussian spatial and temporal pulses profiles, the intensity in a point with space coordinates (x, y, z)

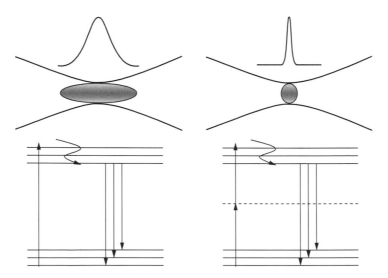

Figure 5.7 Jablonsky diagram comparing the photophysical pathways for single-photon excited fluorescence (bottom left) and two-photon excited fluorescence (bottom right). According to the order of the excitation process, in a focused beam the fluorescence is confined to the focal center of the beam waist in the nonlinear version of fluorescence (top right), whereas it spreads out within a larger volume in the linear process (top left).

at time t is given as:

$$I(x,y,z,t) = I_0 \exp\left[-2\frac{(x^2+y^2)}{W_0^2}\right]\exp\left[-2\frac{Z^2}{Z_0^2}\right]\exp\left[-\left(\frac{t}{\Delta\tau}\right)^2\right] \quad (5.15)$$

where W_0 is the waist beam and $\Delta\tau$ is the time duration of the pulse. At the beam waist (where $z = 0$), the energy per pulse is

$$E = \frac{P}{f} = \iint I dS dt \quad (5.16)$$

where $dS = dxdy$, P is the mean power, and f is the frequency of the pulses. We obtain

$$E(z = 0) = \pi\sqrt{\pi}I_0\frac{W_0^2}{2}\Delta\tau \quad (5.17)$$

Combining Equation 5.17 and Equation 5.16 we finally obtain

$$I_0 = \frac{2}{\pi\sqrt{\pi}}\frac{P}{f}\frac{1}{\Delta\tau W_0^2} \quad (5.18)$$

The maximal flux of incident photons is derived as

$$F_{max} = \frac{I_0\lambda}{hc} = \frac{2}{\pi\sqrt{\pi}}\frac{P}{f}\frac{\lambda}{hc}\frac{1}{\Delta\tau W_0^2} \quad (5.19)$$

Thus Equation 5.19 clearly indicates the shorter the time duration of the pulse and the radius of the beam waist, the more intense the flux of incident photons.

In general, the influence of a signal response that depends nonlinearly on the input intensity is to reduce the effective interaction volume and thus increase the resolution with respect to the linear process. Hence, taking again a Gaussian beam, the interaction volume of an N-photon process decreases by a factor of $\sqrt{N}$, relative to the linear interaction volume at the same wavelength. As an example, in a two-photon excited process, the interaction volume is decreased by a factor of $\sqrt{2}$. In a nonlinear interaction process, since the response volume is confined close around the beam waist, two-photon excited techniques confer 3-D imaging with out-of-focus background rejection similar to a confocal microscope.

Various configurations have been used in multiphoton absorption microscopy. In all cases either laser scanning or specimen scanning is

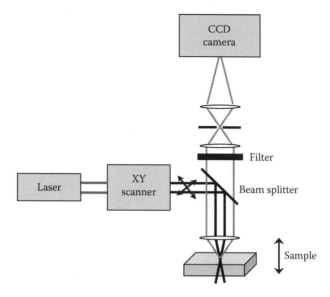

Figure 5.8 General scheme of a two-photon fluorescence microscope. The fluorescence is detected in the backscattering direction and may be imaged on a CCD camera. The sample is scanned in a point excitation mode, using mirror scanning for the X and Y directions and the sample is scanned in the Z direction.

required. This can be accomplished in several ways: adaptation of a confocal microscope retaining the descanned detection channel; use of laser-scanning optics and a specific detection channel without descanning optics and optimized for fluorescence collection efficiency; or use of laser-scanning optics in combination with wide-field charge-coupled device (CCD) detection. A scheme of a typical two-photon microscope setup is depicted in Figure 5.8.

Typically, multiphoton experiments have been performed almost exclusively with near infrared (NIR) lasers, especially Ti:sapphire (700–1100 nm) and Nd:YLF (1047 or 1053 nm). Because water absorption increases significantly at a longer wavelength in going from the visible to NIR region of the spectrum in biological applications, biological damage through heating can be expected (Schönle and Hell 1998). For ultrashort pulses, laser-induced breakdown is a threshold phenomenon. Multiphoton absorption (up to several orders) causes highly localized plasma formation. The rapid expansion of the plasma causes a microexplosion, which results in observable damage. The created damage structure is considered to be either a small change in the refractive index or a vacuum bubble. For transparent media and ~100 fs pulses, as an indicative order of magnitude, the laser-induced breakdown threshold is approximately 60 nJ/pulse. Thus, ample power levels, obtained with shorter pulses, for example, are available for parallel excitation (instead of point excitation) to speed up the image acquisition process.

5.2.2   Two-photon excited second-harmonic generation imaging

Coherent SHG, in contrast to TPEF, does not arise from an absorptive process. Instead, an intense laser field induces a nonlinear polarization of the medium, confined also to the focal center of the beam waist, resulting in the production of a coherent wave at exactly twice the incident frequency. Additionally, the magnitude of the SHG response can be enhanced by tuning the two-photon excitation close to the band gap of a charge-transfer band (Figure 5.9).

We discussed in Chapter 3 that the interaction of collimated beams onto interfaces radiates coherent SHG in well-defined transmission and reflection directions. If we now intend to provide microscopic SHG images, the incident laser beam must be focused on a small spot size and the driving fields can no longer be considered as simple plane waves as before. The structure of the radiating SHG becomes critically dependent on the distribution of the input fields near the focal center. Thus, with focused beams, the phases of the input fields play a crucial role. We have seen in Section 5.1 that in tightly focused beams there is an axial phase shift of the focal field, known as the Gouy phase shift. Since the phase of the excitation beam is phase-delayed by the Gouy shift close to the focal point, the axial effective incident wave vector is diminished as

$$k_\omega^{\textit{effective}} = k_\omega - k_{\textit{Gouy}} \qquad (5.20)$$

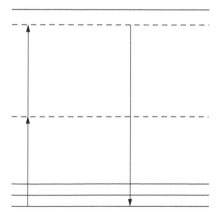

Figure 5.9 Photophysical pathway of resonance-enhanced second-harmonic generation.

where k_ω is the wave vector of a collimated but unfocused beam and k_{Gouy} is the wave vector induced by the Gouy phase shift. As a consequence, when using a tightly focused laser beam, the phase-matching condition for SHG can be revised as

$$\Delta k = [2(k_\omega - k_{Gouy}) - k_{2\omega}] \approx 0 \tag{5.21}$$

The coherence length in the focused beam is then defined by

$$L_c = \frac{\pi}{\Delta k} = \frac{\pi}{2(k_\omega - k_{Gouy}) - k_{2\omega}} \tag{5.22}$$

Conservation of the transfer moment along the axial direction drives a coherent SHG signal within two well-defined off-axis lobes with an angle θ_{peak} that depends on many parameters (Moreaux et al. 2000; Mertz and Moreaux 2001; Figure 5.10). As an indication, for low numerical aperture (NA < 1):

$$\theta_{peak} \approx \theta_{NA}/\sqrt{2} = \arcsin(NA/n)/\sqrt{2} \tag{5.23}$$

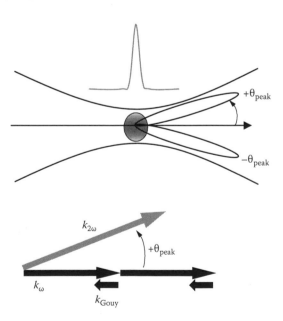

Figure 5.10 Coherent SHG in a tightly focused beam is confined at the beam waist into two well-defined off-axis lobes by phase matching conditions.

Thus the two lobes should always be observed regardless of the NA. However, following Moreaux et al. (2000), at large NA, since the interaction volume decreases (as well as the phase coherence), the angular separation of the two lobes increases; but there is a broadening of the lobes themselves. In opposition, at small NA, the two lobes should be better seen since their distribution is sharper.

Coherent techniques require a different optical setup than incoherent techniques like TPEF. Because of their complementarities, SHG and TPEF imaging are currently combined in a compact setup.

Since forward SHG is more intense than backward SHG because of the coherent nature of the process, SHG is detected in the forward direction, whereas TPEF imaging is collected in the backscattering geometry (Campagnola 1999; Squier and Müller 2001; Barzda et al. 2005; Strupler et al. 2007). However, it has been shown that forward and backward SHG images are different and give complementary structural information (Pfeffer et al. 2007; Figure 5.11). Here again, as in TPEF imaging, SHG imaging have been performed almost exclusively with NIR lasers, especially Ti:sapphire (700-1100 nm) and Nd:YLF (1047 or 1053 nm).

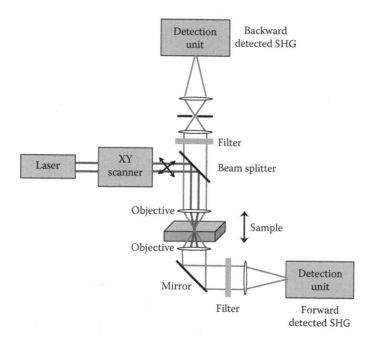

Figure 5.11 Optical schematic of a SHG microscope with a forward and backward detection unit for imaging. The sample is scanned in a point excitation mode, using mirror scanning for the X and Y directions and the sample is scanned in the Z direction.

5.3   Applications

Applications of nonlinear microscopy are numerous but most of them focus on biological applications (Mertz 2004). Multiphoton fluorescence microscopy can elegantly be combined with SHG imaging providing unique but complementary information about biological structures of noncentrosymmetric molecular organization.

Of particular relevance to biological imaging are collagen, microtubules, and myosin. For instance, Campagnola et al. (1999) were able to produce live cell images based on SHG imaging and they also demonstrated the enhancement of SHG due to chirality. Indeed, since this coherent technique requires an environment without a center of symmetry, it is very sensible to any contrast modification at any interfacial region that produces an SHG signal. Note that in the case of cells, although the medium is isotropic, the membranes are locally asymmetric for the beams since their typical size (a few tenth of micrometer) is larger than the wavelength.

Over the last decades, applications in the studies of equilibrium properties, absolute orientations of molecules in interfacial regions, and studies of dynamical processes have been done. Because of the inherent asymmetry of lipid bilayers, both intracellular organelle and plasma membranes are suitable samples for probing with this methodology. Bouevitch et al. (1993) have imaged SHG signals in a model membrane stained with a voltage-sensitive dye on a hemi-spherical bilayer apparatus. Pfeffer et al. (2007) have evidenced differences in the image features of the forward detection scheme that show the fibrillar nature of fascia collagen matrix. In contrast, the backward features reflect the submicron architecture. The SHG radiation pattern is mainly determined by the phase-matching condition as we have seen in the previous section. Therefore, previous theoretical and experimental works have revealed that objects with the axial size on the order of the second-harmonic wavelength exhibit forward directed SHG, while objects with an axial size less than $\lambda/10$ (approximately 40 nm) are estimated to produce nearly equal backward and forward SHG signals (Moreaux et al. 2000; Mertz and Moreaux 2001; Williams et al. 2005). Then, the ratio of the forward/backward signal (F/B) is an indication of the axial size of the scatterer in the focal volume. As observed also by Strupler et al. (2007), the main advantage of SHG microscopy is its unique capability to provide 3-D images of the organization of collagen fibers with micrometer resolution. Additionally, the high specificity of the SHG signal for fibrillar collagens results in an intrinsically small background in the images that enables sensitive measurements. Also, from the polarization dependence of SHG intensity it is possible to gain orientational information, like an estimation of the helical angles of myosin and collagen (Tiaho et al. 2007).

Finally, as one uses fluorophores in two-photon absorption, it is possible to introduce highly polarizable molecules as SHG markers. Keeping in mind that the magnitude of the SHG response can be resonance enhanced when the energy of the two incident photons overlaps with an electronic absorption band (see Figure 5.9), the total SHG response is

$$\chi^{(2)}_{total} = \chi^{(2)}_{nonres} + \chi^{(2)}_{res} \tag{5.24}$$

Depending on the specific properties of the chromophores or tuning the frequency of the incident laser beam, the resonant contribution from the SHG markers can dominate and an enhancement of one to two orders of magnitude can be expected. However, a major constraint of SHG is the requirement of non-centrosymmetric assemblies of dipolar moieties in the electric dipole approximation. Of course, specific chiral molecular markers can be also used. As a final potential application not yet explored as far as we know, photo- or acido-chromic molecules with sufficient NLO contrast could be used as specific photo- or acido-switchable SHG markers (see, for example, Mançois et al. 2007).

In the area of electronic and/or photonic materials or devices, SHG imaging can be a powerful tool for visualizing embedded electric fields inside materials, based on the electric field induced second harmonic (EFISH) contribution (see, for example, Chapter 3, Section 3.3.2). Manaka et al. (2006, 2007) have determined the electric field distribution in organic field effect transistor (FET) devices by optical SHG imaging on the basis of the EFISH contribution. Also, EFISH contributions have been determined from polarized SHG imaging of the cross-section of a thermally poled sodium borophosphate glass plate (Rodriguez et al. 2006; Figure 5.12). The

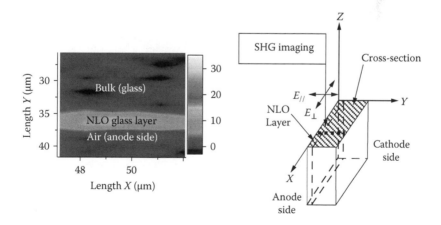

Figure 5.12 Polarized ($E_{//}$) SHG image of the cross-section of a thermally poled glass plate, obtained with a NIR 100× objective (NA = 0.5). The acquisition time per pixel was 10 s. The polarized ($E_\perp$) SHG image gave no signal indicating that the electric field is along the Y direction as expected.

polarized SHG images (15×15 μm^2) have been obtained in the reflection mode, that is, in the backward direction. The SHG-active layer (~4 μm thickness) obtained from a thermal poling treatment (see Chapter 3, Section 3.3.2, or Dussauze et al. 2005 for additional details) is clearly observed with an axial resolution of ~1.0 μm. As expected, an incident polarization ($E_{//}$), parallel to the embedded electric field along the Y direction, gave a SHG signal, whereas a perpendicular incident excitation ($E_{\perp}$) gave no signal. In this original setup, micro-hyper-Raman spectra in the backscattering mode coupled with backward-SHG imaging with an axial resolution of ~1 μm has recently been obtained, which opens a new and large field of applications.

References

Barad, Y., et al., *Appl. Phys. Lett.*, 70 (1997), 922.

Barzda, V., C. Greenhalgh, J. aus der Au, S. Elmore, J.H.G.M. van Beek, and J. Squier, *Opt. Express*, 13 (2005), 8263.

Born, M., and E. Wolf, *Principles of Optics*, 4th ed., New York: Pergamon, 1970.

Bouevitch, O., A. Lewis, I. Pinevsky, J.P. Wuskell, and L.M. Loew, *Biophys. J.*, 65 (1993), 672.

Bruneel, J.L., J.C. Lassègues, and C. Sourisseau, *J. Raman Spectrosc.*, 33 (2002), 815.

Campagnola, P.J., M. Wei, A. Lewis, and L. Loew, *Biophys. J.*, 77 (1999), 3341.

Clark, R.H.J., and R.E. Hester, eds., *Advances in Nonlinear Spectroscopy*, New York: Wiley, 1984.

Denk, W., J.H. Strickler, and W.W. Webb, *Science*, 248 (1990), 73.

Diaspro, A., ed., *Confocal and Two-Photon Microscopy: Foundations, Applications, and Advances*, New York: Wiley-Liss, 2002.

Dussauze, M., E. Fargin, M. Lahaye, V. Rodriguez, and F. Adamietz, *Opt. Express*, 13 (2005), 4064.

Everall, N.J., *Appl. Spectrosc.*, 54 (2000a), 773.

Everall, N.J., *Appl. Spectrosc.*, 54 (2000b), 1515.

Ippolito, S.B., B.B. Goldberg, and M.S. Ünlü, *J. Appl. Phys.*, 97 (2005), 053105.

Kneipp, J., H. Kneipp, B. Wittig, and K. Kneipp, *Nano Lett.*, 7 (2007), 2819.

Kneipp, K., H. Kneipp, I. Itzkan, R.R. Dasari, and M.S. Feld, *Chem. Rev.*, 99 (1999), 2957.

Kneipp, K., H. Kneipp, and J. Kneipp, *Acc. Chem. Res.*, 39 (2006), 443.

Kohen, E., and J.G. Hirschberg, *Cell Structure and Function by Microspectrofluorometry*, San Diego, CA: Academic, 1989.

Lewis, I.R., and H.G.M. Edwards, eds., *Handbook of Raman Spectroscopy*, New York: Marcel Dekker, 2001.

Manaka, T., E. Lim, R. Tamura, D. Yamada, and M. Iwamoto, *Appl. Phys. Lett.*, 89 (2006), 072113.

Manaka, T., M. Nakao, D. Yamada, E. Lim, and M. Iwamoto, *Opt. Express*, 15 (2007), 15964.

Mançois, F., L. Sanguinet, J.L. Pozzo, M. Guillaume, B. Champagne, V. Rodriguez, F. Adamietz, L. Ducasse, and F. Castet, *J. Phys. Chem. B*, 111 (2007), 9795.

Mertz, J., *Curr. Opinion Neurobiol.*, 14 (2004), 610.

Mertz, J., and L. Moreaux, *Opt. Comm.*, 196 (2001), 325.

Moreaux, L., O. Sandre, and J. Mertz, *J. Opt. Soc. Am. B*, 17 (2000), 1685.

Müller, M., et al., *J. Microcs.*, 191 (1998), 266.

Pawley, J.B., *Handbook of Biological Confocal Microscopy*, New York: Plenum, 1995.

Pfeffer, C.P., B.R. Olsen, and F. Légaré, *Opt. Express*, 15 (2007), 7296.

Rodriguez, V., D. Talaga, F. Adamietz, J.L. Bruneel, M. Couzi, *Chem. Phys. Lett.*, 431 (2006), 190.

Schönle, A., and S.W. Hell, *Opt. Lett.*, 23 (1998), 325.

Sourisseau, C., *Chem. Rev.*, 104 (2004), 3851.

Squier, J., and M. Müller, *Rev. Scient. Inst.*, 72 (2001), 2855.

Strupler, M., A.-M. Pena, M. Hernest, P.-L. Tharaux, J.-L. Martin, E. Beaurepaire, and M.-C. Schanne-Klein, *Opt. Express*, 15 (2007), 4054.

Svoboda, K., et al., *Nature*, 385 (1997), 161.

Tiaho, F., G. Recher, and D. Rouède, *Opt. Express*, 15 (2007), 12286.

Williams, R.M., W.R. Zipfel, and W.W. Webb, *Biophys. J.*, 88 (2005), 1377.

Appendix

We chose to use the Gaussian unit system in this book because relations describing electromagnetic phenomena are clearer and much simpler when defined in the Gaussian system. For example, P, E, H, M, D, and B all have the same units, whereas this is not the case in the SI unit system. This can be clearly seen from the Maxwell equations and the constitutive relations:

$$\nabla.D = 4\pi\rho \qquad \nabla.B = 0$$

$$\nabla\times E = -\frac{1}{c}\frac{\partial B}{\partial t} \qquad \nabla\times B = \frac{1}{c}\frac{\partial D}{\partial t} + \frac{4\pi}{c}J \qquad \text{(A.1)}$$

$$D = E + 4\pi P \qquad B = H + 4\pi M$$

In the SI definition of these equations this is clearly not the case, since one has to include permeability (μ_0) and permittivity (ε_0) of vacuum:

$$\nabla.D = \rho \qquad \nabla.B = 0$$

$$\nabla\times E = -\frac{\partial B}{\partial t} \qquad \nabla\times H = \frac{\partial D}{\partial t} + J \qquad \text{(A.2)}$$

$$D = \varepsilon_0 E + P \qquad B = \mu_0 H + M$$

However, since the SI unit system is now generally accepted as the unit system of choice, we included this appendix that briefly describes conversion relations between the Gaussian and SI unit systems.

In the Gaussian system, polarization is defined as:

$$P = \chi^{(1)}E + \chi^{(2)}EE + \cdots \qquad \text{(A.3)}$$

Since polarization and electric field are both expressed in statvolt/cm, it is clear that the linear susceptibility is dimensionless and $\chi^{(2)}$ is given in cm/statvolt. Note that 1 statvolt is 299.792458 V. In most cases, however, units are usually not given and it is simply stated that the value is given in electrostatic units (esu).

In the SI unit system, there are two conventions that define polarization slightly different. In the first convention, polarization is defined as:

$$P = \varepsilon_0 (\chi^{(1)} E + \chi^{(2)} EE + \cdots) \qquad (A.4)$$

whereas in the second convention it is defined as

$$P = \varepsilon_0 \chi^{(1)} E + \chi^{(2)} EE + \cdots \qquad (A.5)$$

Since polarization is expressed in C/m² and electric field in V/m, the linear susceptibility is dimensionless in both conventions and the second-order susceptibility is expressed in m/V (convention 1) or C/V² (convention 2).

Similar arguments can be used to derive the units of polarizability and hyperpolarizability. However, since susceptibilities are basically polarizabilities per unit volume, one simply has to multiply the units of the respective susceptibilities with cm³ (Gaussian) or m³ (SI).

Although it is simple to derive the units in each unit system, it is more difficult to derive conversion factors between both unit systems since Maxwell's equations and the constitutive relations are defined differently in both unit systems. It is convenient to start from a quantity that is the same in all unit systems, such as the refractive index. In Chapter 1, the refractive index (n) was introduced via the relation (Gaussian system):

$$D = E + 4\pi P = (1 + 4\pi\chi^{(1)})E = n^2 E \qquad (A.6)$$

Since nonlinear optical interactions are just small corrections to the refractive index, we may also write:

$$D = E(1 + 4\pi \chi^{(1)} + 4\pi\chi^{(2)} E) = n^2 E \qquad (A.7)$$

In the SI unit system these relations are defined as:

$$D = \varepsilon_0 E(1 + \chi^{(1)} + \chi^{(2)}E) = \varepsilon_0 n^2 E \quad \text{(convention 1)} \tag{A.8}$$

$$D = \varepsilon_0 E\left(1 + \chi^{(1)} + \frac{\chi^{(2)}}{\varepsilon_0}E\right) = \varepsilon_0 n^2 E \quad \text{(convention 2)} \tag{A.9}$$

Knowing that $E(SI) \cong 3 \times 10^4 E(cgs)$, the conversion factors are given by:

$$\chi^{(1)}(SI, convention\ 1, 2) = 4\pi\chi^{(1)}(cgs) \tag{A.10}$$

$$\chi^{(2)}(SI, convention\ 1) = \frac{4\pi\chi^{(2)}(cgs)}{3 \times 10^4} \tag{A.11}$$

$$\chi^{(2)}(SI, convention\ 2) = \frac{4\pi\varepsilon_0\chi^{(2)}(cgs)}{3 \times 10^4} \tag{A.12}$$

Index

Printed and bound by CPI Group (UK) Ltd, Croydon, CR0 4YY

21/10/2024

01777083-0006